Roe and Roe Gastronomy

Ole G. Mouritsen • Klavs Styrbæk

Roe and Roe Gastronomy

Ole G. Mouritsen
Odense M, Denmark

Klavs Styrbæk
STYRBÆKS
Odense N, Denmark

ISBN 978-3-032-13141-6 ISBN 978-3-032-13142-3 (eBook)
https://doi.org/10.1007/978-3-032-13142-3

Based on the authors' own translation of the Danish edition: Rogn – Meget Mere end Caviar. Gyldendal, Copenhagen (2023) 246pp.

Cover: Salmon roe. Photography by Jonas Drotner Mouritsen

Illustrations by Jonas Drotner Mouritsen

This Springer imprint is published by the registered company Springer Nature Switzerland AG
The registered company address is: Gewerbestrasse 11, 6330 Cham, Switzerland

Preface

This book is based on the puzzling observation that, apart from a few different types of roe, such as cod roe, lumpfish roe, and sturgeon caviar, roe from most other edible fish, and not to speak about roe from other marine animals such as crustaceans, mussels, and other mollusks, play only a very minor role in Western cuisine. This is remarkable for many reasons.

Firstly, coastal nations have a rich tradition of eating food from the sea. Secondly, roe are a healthy and nutritious food source with large amounts of important omega-3 fatty acids. Thirdly, roe and eggs from marine organisms are found in a bounty of different colors, sizes, flavors, and textures, so there are rich variations for gastronomic uses. And finally, and this is the main point of the book, roe, thanks to its strong umami taste, can help us add taste to a more plant-rich and sustainable cuisine.

In some food cultures, such as in East Asia and Southern Europe, roe play a significant role in everyday cooking and as a gastronomic delicacy. This could also be the case in other countries, but this would require that we start to ask for fresh roe from the fishmonger and also learn to use the roe in the kitchen. Apart from a few different types of roe, most of which are seasonal or canned, consumers will go to the fishmonger in vain for roe. In most cases, the roe are cut off from the fresh fish before it reaches the consumer. The cut roe then often goes to waste or is used in feed for breeding fish and livestock. This is not a sustainable use of a valuable food source.

On this backdrop, we have set out to write a book about roe to open the consumers' eyes to this overlooked delicacy from the sea. Our ambition has been to combine knowledge of biology, cultural history, culinary science, and gastronomy with good food craftsmanship in a book that provides a popular description of what roe are and how we can use them in our kitchen in a sustainable way with respect for nature and with a focus on taste and texture, not least in a plant-forward cuisine.

Roe are the eggs from fish, crustaceans, and mollusks. The eggs contain the germ of the next generation along with a lunch bag of nutrients for the offspring. Roe are also food for other animals in and around the sea, as well as for humans. Roe can be eggs from sturgeon but is also much more than caviar. The book therefore also deals with the roe we call *k*aviar from other fish species such as flatfish, cod, mullet, halibut, capelin, and herring as well as from crabs, shrimp, lobster, squid, and sea urchin. In addition, kaviar substitutes made from fish meat and seaweed are discussed.

A number of books have been published about caviar from sturgeon, but strangely enough, there are no cookbooks and gastronomic works dedicated to roe in general. There is also surprisingly little in the international scientific literature on the sensory and physicochemical properties of roe and fish eggs in relation to taste, aroma, and texture, even though these factors are important for product development, consumer preferences, and quality control. The present book seeks to remedy some of these shortcomings by collecting, synthesizing, and communicating the existing knowledge, both from gastronomy and science, in addition to placing roe in the context of culinary craft and art.

During the collection of quantitative data for the book, it became clear that there is a need to produce reliable scientific data on the taste of roe in particular. The book project has thus stimulated a new research program at the University of Copenhagen, supported by the Carlsberg Foundation and the European Committee for Umami, where we have set out to measure specifically umami taste substances in roe from a wide range of marine animals. When collecting roe for this research project, we realized that it is not straightforward to get fishmongers to provide fresh roe. Roe are simply not in demand by consumers. Fortunately, we succeeded in collecting roe for 30–40 species, and scientific articles are in the process of being published with the results of our studies. Some of the preliminary results of our work have already found a home in the present book.

The book is a product of a collaboration between a chef and a scientist who have ventured into the kitchen together to learn more about roe by tasting, experimenting, and showing an interest in combining curiosity and knowledge with good cooking. The book illustrates the different uses of many types of roe in both classic and newly interpreted dishes as well as completely novel preparations. The book is therefore aimed at a broad audience who is interested in the world and the part of the world we eat.

During the work on researching for the book, we have received help and guidance from a number of colleagues and good contacts. Special thanks are due to

- Kasper Styrbæk for testing recipes and dishes
- Poul Rasmussen for countless conversations about fish, roe, and smoking of cod roe
- Charlotte Vinter Schmidt and Karsten Olsen for scientific collaboration on roe
- Sofie Raae Søndergaard for collecting and preparing roe samples for analysis
- Jacob Marsing-Rossini for supplying caviar and general information about caviar production
- Morten Priess, Henning Priess, and Kurt Mogensen, AquaPri, for information about the production of trout roe and for hospitality during a visit to the factory in Aarøsund

- Keiichiro Hayakawa, Mitsui & Co., Tokyo, for a tour and explanations about Danish production of *sujiko* for the Japanese market
- Jesper Kold Sørensen, Ronnie Nielsen, and Brian Gert, Amanda Seafoods A/S, for information about cod roe and for hospitality during a visit to Amanda in Frederikshavn
- Johannes Palsson and Benjamin Bosse, Scandic Pelagic, for information on herring roe production and for hospitality during a visit to Scandic Pelagic in Skagen
- Bastien Debeuf, Groupe KAVIAR, Saint-Fort-sur-Gironde, for information on sturgeon aquaculture and caviar production
- Ignacio Alba Alejandre and Laura Cobos Cano, Caviar de Riofrío, for a tour of Riofrío and information on caviar production
- Paolo Bronzi, World Sturgeon Conservation Society, for data on world caviar production
- André Magalhães, Taberna Rua de Las Flores, Lisbon, for a sample of octopus roe *bottarga* and the corresponding recipe
- Rasmus Munk from Restaurant Alchemist for king crab roe
- Ditte Ankjærgaard, Nordic Snails, for the supply of snail eggs
- Albanigades Fisk og Vildt, HAV i Torvehallerne, Fiskerikajen, Sømunken, and Hilbert Christiansen Fisk and Vildt for supplying samples of fresh roe
- Jens Møller and Jens Christian Møller for samples of Cavi-art
- Birger Brix for black lobster roe
- Susane Røntved for information about herring roe
- Minaka One for information about the use of roe in Japanese cuisine
- Henrik Carl, Statens Naturhistoriske Museum, for information about fish
- Just León, Pescaviar, for information about the herring-based caviar substitutes Avruga and Moluga
- Carolina Camacho, Portuguese Institute of Sea and Atmosphere, for information about ingredients in sea urchin roe
- Anna Górka for information about snail eggs
- Bent Brandt A/S for lending service for photo sessions

The book was originally written and published in Danish, the authors' mother tongue, in 2023. The translation to English and the adaptation of the book to an international readership began for one of us (Ole) during a scholarship stay in September 2025 at the convent San Cataldo in the little town Scale on the Amalfi Coast in Southern Italy. The tranquility of working in the thousand-year-old convent, combined with the stay in the beautiful ancient cultural landscape, which is on the UNESCO World Heritage List, was a perfect setting to immerse oneself in the fascinating world of roe and especially their culinary uses.

A number of people and organizations have kindly made illustrative material available for the book. A list of these can be found at the back of the book. A book of this type with many illustrations and photos could not be produced without financial support. The publication was supported by VELUX FONDEN and Albani Fonden. The authors thank the foundations for their support.

The research behind some of the new scientific results presented in the book on the umami potential of roe was supported by a grant from the Carlsberg Foundation. Jonas Drotner Mouritsen has participated in the project from the very beginning as a designer, photographer, and layout designer. We hope that the attentive reader will be able to see that the collaboration between a chef, a scientist, and a designer has led to a book that is holistic in more than one sense.

Odense, Denmark
October 2025

Ole G. Mouritsen
Klavs Styrbæk

Contents

About the Authors

Ole G. Mouritsen, PhD DSc dr.h.c. FRS-DK, is a research scientist and Professor Emeritus of gastrophysics and culinary food innovation at the University of Copenhagen, Denmark. He holds an honorary professorship with the University of Melbourne, Australia. His scientific work has focused on basic sciences and their applications within the fields of physical chemistry, biophysics, biomedicine, and food. He is the recipient of numerous prizes for his research work and for research communication. His extensive list of publications includes six monographs co-authored with chef Klavs Styrbæk, which integrate scientific insights with culinary perspectives and have been nominated several times for Gourmand Best in the World Awards. Currently, he is president of The Danish Gastronomical Academy and past director and founder of the national Danish center Taste for Life. For many years, he has been fascinated with the Japanese culinary arts and explaining the extent to which their techniques and taste elements can be adapted for the Western kitchen. In recognition of his efforts, he was appointed in 2016 as a Japanese Cuisine Goodwill Ambassador by the Japanese Ministry of Agriculture, Forestry, and Fisheries, and in 2017 the Japanese Emperor bestowed upon him The Order of the Rising Sun, Gold Rays with Neck Ribbon, *Kyokujitsu chūjūshō* 旭日中綬章.

Klavs Styrbæk is a professional chef who owns and operates STYRBÆKS together with his wife, Pia. By combining a high standard of craftsmanship, sparked by curiosity-driven enthusiasm, he has created a gourmet center where people can enjoy excellent food and where they can come to learn and take their culinary skills to a whole new level. He is particularly enthusiastic about seeking out unique, local raw ingredients that are incorporated into new taste adventures or used to revisit traditional Danish recipes that might otherwise be forgotten. This delicate balance between innovation and renewal is demonstrated in his award-winning cookbook *Mormors mad* (*Grandmother's Food*, 2006), which was honored with a special jury prize at the Gourmand World Cookbook Awards in 2007. In 2008 and 2019, he was awarded an honorary diploma for excellence in the culinary arts by the Danish Gastronomical Academy. Many of the recipes that appear in the books co-authored with Ole originated in the test-kitchens at STYRBÆKS.

List of Recipes

In many of the book's recipes, one type of roe can replace another, and you can therefore use the same recipes in different seasons, regardless of what type of roe you have available

Chapter 1
Introduction: Roe and Food Culture

All food cultures that are connected to the sea, rivers, or lakes have a tradition of eating roe from fish, in particular, but also to some extent from mollusks and crustaceans. Generally, it is especially roe with large eggs, e.g., from salmon and sturgeon, that have been associated with gastronomic value, whereas the fine-grained roe from, e.g., flatfish have been less respected and sometimes even wasted or have ended up as feed for other fish or livestock.

The most famous roe product is undoubtedly caviar from sturgeon, but the world's consumption of roe from other fish species is much greater. In the Western world, it is mainly salted roe, possibly dried roe, and the most popular products in Middle Eastern and Far Eastern countries are salted and fermented types of roe. In these regions of the world, roe from many different fish and other marine animals are consumed, and roe are considered a great delicacy.

Roe consists of a collection of individual eggs, typically enclosed in a roe sac (cf. Fig. 1.1). Because the eggs contain the germ for the next generation, they are filled with valuable nutrients, especially unsaturated fats, proteins, amino acids, and nucleic acids. With this nutritious food package, the egg cell itself, which is contained inside the egg, can develop and grow into a new individual, which hatches from the egg and can then later seek its own food when the food package is used up. It is the content of this food package, especially fats, certain aromas, and umami-tasting substances, that is the reason why roe have been a sought-after food around the world, both in the daily diet and as a gastronomic specialty. In addition, the individual eggs have a very special mouthfeel (texture), which is determined by whether the roe are eaten raw or prepared, typically by salting and cooking.

Most food cultures with access to the sea have old traditions of eating roe from local fish and with seasonal variations. In some cultures, these traditions have been maintained, whereas in others where a whole fish seldom reaches the consumer, roe are a less common and familiar food item and mostly consumed in the form of preserved loose eggs used to garnish dishes such as open sandwiches, fish dishes, and especially egg dishes. Also, the selection of roe types used is very narrow.

O. G. Mouritsen, K. Styrbæk, *Roe and Roe Gastronomy*,
https://doi.org/10.1007/978-3-032-13142-3_1

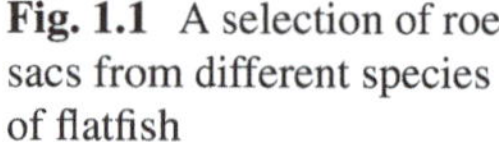

Fig. 1.1 A selection of roe sacs from different species of flatfish

But all fish carry roe, and it is not only caviar from sturgeon that is tasty, nutritious, and a fine delicacy. It could just as well be roe from common edible fish such as flatfish, garfish, mackerel, and herring. However, consumers rarely see this roe because it is cut off before the fish reaches the consumer. We have forgotten to ask for roe, and we have disregarded the food culture around preparing roe in the kitchen. This book seeks to remedy this situation.

We do not eat food, at least not in the long run, if we do not like the taste of it. But here roe have something very special to offer, because it provides for an intense umami taste, and the individual eggs often have exciting textures such as crispy, crunchy, or melting. All people like umami taste, because our species is evolutionarily and genetically developed to strive for this particular taste. We know that we should eat more plant-based foods for our own health and the health of the planet, but green foods like vegetables and legumes often lack umami. Adding more roe and other marine food to our diet can therefore help us to choose a more plant-rich diet that is more sustainable and more climate-friendly. All of this is explained in this book.

With some selected recipes, we will show how all kinds of roe can be used to add taste and mouthfeel to simple dishes that anyone can make in their own kitchen. It is not difficult. Some recipes are completely classic, others are new interpretations of classic dishes with roe, and others are new and exciting dishes. The main line in all the recipes is that taste comes first. To whet the appetite, Fig. 1.2 shows a green salad with roe and lots of umami taste.

Recipe 1.1 Salad with Roe and Wheat Kernels
Serves 4

100–150 g roe as kaviar from, e.g. salmon, herring, or lumpfish as well as smaller roe sacs
200 g wheat kernels
Salt and pepper
2 avocados
1 cucumber
3–4 unpeeled limes
1 bunch spring onions
2 tbsp olive oil
2 slices sandwich bread
1 tbsp sesame oil
Fresh coriander or parsley

1. Cook the wheat kernels according to the instructions on the package.
2. Pour off the cooking water and marinate the kernels while they are still warm, with the grated peel and the juice of 2–3 limes. Season with salt and pepper.
3. Peel the cucumber, split it lengthwise, remove the seeds, and cut into cubes.
4. Remove the peel and stone from the avocados, cut them into cubes, place them in a bowl, drizzle with lime juice, and sprinkle with a little salt.
5. Cut the spring onions into thin rings and the last lime into wedges.
6. Cut the crusts off the sandwich breads and cut the rest into small cubes.
7. Toast the bread cubes into croutons on a dry pan until they are light brown, add sesame oil, toast them quickly, and place them on absorbent paper.
8. Roll the roe sacs in flour, season, and fry them with butter in a pan.
9. Mix wheat grains with cucumber, avocado, olive oil, and spring onions, and place the salad in a wide bowl.
10. Spread the roe beautifully over the salad, cut coriander leaves on top, add lime wedges, and sprinkle with bread croutons.

Some roe are best prepared in salads, for example the whole smaller roe sacs of flatfish, which can be dusted in flour and fried in butter. Other types of roe such as *löjrom/sikrom*, lumpfish roe, salmon roe, sea urchin roe, or flying fish roe can be used as they are, fresh and lightly salted or bought in jars. Others, such as herring roe, can be used both prepared and fresh, after they have been frozen for at least 24 hours.

Fig. 1.2 Salad with different types of roe, wheat grains, avocado, lime, coriander, and sesame bread croutons

1.1 The Green Transition

The so-called green transition has become a central issue in politics, production, transport, energy, food, health, and climate. The starting point is the now very clear and negative impact of human activities on the state of our planet, not least manifested in climate change. Our era has even been called the Anthropocene epoch, i.e., the era in which human activities have left a decisive and lasting impact on the state of the planet.

It is now indisputable that food production is a significant cause of changes in the Earth's ecosystems, including climate change. This applies both to agriculture, which is responsible for 40% of land use, 30% of greenhouse gas emissions, and 70% of the use of fresh water, and to fisheries, where 60% of the ocean's fish stocks are fully exploited and 30% overfished, in addition to the fact that catches have been declining in recent decades. The implications of this type of food production are a huge loss of biodiversity, overuse of fresh water, destruction of whole ecosystems, and the emission of excess nutrients and greenhouse gases. The Earth's cycles of essential elements such as carbon, phosphorus, and nitrogen have been disrupted. This food system is not sustainable.

Furthermore, even though this unsustainable food system produces enough calories for the present population of the Earth, especially when food waste is taken into account, and even though resources are unequally distributed and 820 million people are hungry, diet-related diseases, such as diabetes, cardiovascular disease, high blood pressure, cancer, etc., have increased at a galloping pace. The present diet is therefore neither sustainable nor healthy on a global scale. If we add to this bleak description that the Earth's population is growing rapidly and will approach 9.8 billion people by 2050, it is clear that it is critical to change the way we produce and consume food. It is this change that is central to the green transition.

Based on extensive calculations and analyses, an international commission of experts from 16 different countries (the EAT–Lancet Commission on healthy diets from sustainable food systems) presented in 2019 a recommendation in a report on Earth's food systems for a so-called planetary diet or menu that is healthy, nutritious, and sustainable. This diet consists predominantly of vegetables, fruits, whole grains, legumes, nuts, and unsaturated fats, with only moderate or small amounts of fish and poultry and no or very little red meat, processed meat, added sugar, refined cereals, and starchy vegetables.

The specific recommendation for a planetary diet prescribes 300 g of vegetables and 200 g of fruit per day, corresponding to five portions. In addition, it includes around 230 g of whole grains (rice, wheat, corn), corresponding to a maximum of 60% of the calorie intake, along with 50 g of starchy vegetables (e.g., potatoes and cassava). The report asserts that with this recommendation it should be possible to meet the UN Sustainable Development Goals regarding a healthy diet that is produced and consumed sustainably. However, the calculations also show that this solution is fragile, and even a small increase in the consumption of red meat and dairy products can have disastrous consequences.

1.2 When Blue Is Green

Fish are an important component of the planetary menu and thus the green transition, because it is from fish and other food from the sea that we can get the absolutely necessary super-unsaturated fats that are so crucial for our health. The planetary menu ought also to include shellfish, mollusks (such as mussels and cephalopods), and seaweed from the sea, as all of these organisms are also sources of super-unsaturated fats. Importantly, seaweed and mussels are particularly sustainable, because they are found at the bottom of the Earth's food web and thus put the least possible stress on the use of resources and the Earth's climate. Plants do not have these super-unsaturated fats. The blue foods from the sea, including roe, thus become important for the green transition for nutritional reasons alone. But then there is the taste.

1.3 Taste Comes First

Seafood is important not only for its nutritional value but also because marine foods can add the necessary umami taste to a diet that is more plant-rich. To understand how important this is, we can start by noting that most people find it difficult to cut down on meat and eat 500–600 g of vegetables and fruit a day and that this is because plants have little umami taste.

Humans are meat eaters and have been so for at least 2 million years, especially after our ancestors began using fire to cook food 1.9 million years ago. A key taste of meat is umami, which is due to the glutamate content but also free nucleotides from muscles. We are evolutionarily and genetically conditioned to crave the umami taste because it has led us to foods with a high protein content and a lot of energy, which is good for the survival and reproduction of our species.

Combining the fact highlighted above with the fact that plants (and here we should exclude ripe fruits) have only a little umami taste, we find the reason why most people have difficulty cutting down on meat and eating more greens, e.g., vegetables. To understand what the solution to this problem could be, it is important to note that it is not meat as such that we crave but the taste of meat, i.e., umami. The way to get more greens in the diet is therefore to find ways in which greens can obtain the "missing" umami taste. The solution is straightforward: we can either add umami taste to green dishes using other umami-rich ingredients, or we can release the greens' own potential to develop umami taste through fermentation. Taste is therefore an important key to promoting the green transition. And this is where roe can play a useful role.

1.4 The Role of Roe in the Green Transition

In this book, we will describe how roe from various organisms such as fish, shellfish, crustaceans, mollusks, and echinoderms can be a source of umami and so-called umami synergy. We will show through selected recipes how little roe is needed to give umami and make green dishes more palatable. By preparing the roe in a special way, you can even enhance the umami taste, e.g., the umami taste in dried and matured roe is ten times greater than in fresh roe. By eating roe from fish, rather than cutting off the roe and letting it end up as animal feed or food waste, we can promote the sustainability of our diet from the sea.

We will therefore advocate for a greater sphere of the use of roe from various marine animals, not least those fish where we only have a tradition of eating the meat, and should start using the roe as it is or as an (umami) flavoring for green dishes. We are of the opinion that by putting focus on adding umami taste to greens with a little help from the animal kingdom, in this case roe, we can help many more people eat a more plant-rich diet without compromising taste. We believe that such a flexitarian approach on every plate could work well for as many people as possible.

1.5 What About Sustainability?

Eating roe immediately raises the question of whether we can justify eating the offspring of animals and, if we do, how this can be done in the most gentle and sustainable way. Many of the ocean's fish stocks are under great pressure, several species are endangered, and overfishing has meant that many fish do not grow to a size where they produce a large amount of roe. This is particularly a problem for large fish such as some sturgeons, which may take 18–20 years to mature to produce roe. There is no simple answer to this question. In some cases, sustainable solutions are sought by farming fish in aquaculture for roe production, so that wild stocks are not stressed. Other approaches consist of harvesting the roe by stroking the live fish without slaughtering it or by surgically removing the roe.

The way we will approach sustainability in this book is to show that roe in small quantities can be used as a kind of "seasoning" that can help us eat a greener and more plant-rich diet. This is especially about providing the greens with the umami taste that they lack by nature. In addition, we will provide recipes on how to make better use of roe and stimulate an appetite for the roe that are found in large quantities in common edible fish, such as cod, herring, and flatfish, and which either go to waste in the fishing industry and home kitchens or are inappropriately prepared and used. We will also discuss how the often-overlooked delights of crustacean roe can act as flavor bombs in sauces and dressings.

Can Roe Be Organic?

In principle, wild, free-swimming fish, and thus their roe, cannot be organic in the strict sense, as it is impossible to know the status of their food and the water of their habitat. Only fish that have been raised under controlled conditions may be classified as organic. Organic roe from farmed trout and sturgeon are available on the market. The sustainability of wild-caught fish is monitored by an international nonprofit organization, the Marine Stewardship Council (MSC), which issues the MSC-Certified Sustainable Fisheries certificate. The MSC operates on a basis of sustainability that safeguards fish stocks, jobs in the fishery, and consideration for the marine environment.

1.6 Different Roe Products

Roe products are usually divided into three main types: whole roe sacs, loose eggs, and pastes or spreads. Roe are rarely eaten completely raw and almost always involve salting and cooking and in some cases drying, fermentation, and various forms of preservation. The best-known products are loose-egg products like kaviar and caviar, and there are also a wide range of substitute products for roe and fish eggs.

1.6.1 *Kaviar and Caviar*

When the eggs are freed and detached from the roe sac for consumption and typically salted and preserved, the loose fish eggs are called kaviar (cf. Fig. 1.3). If

Fig. 1.3 Selection of different types of fish roe in the form of loose eggs (kaviar)

they are eggs from sturgeon, this form of kaviar may be called caviar according to the *Codex Alimentarius*, and all other types of roe may not. However, there are different rules in North America and the EU for the use of the terms kaviar, caviar, and kaviar substitute. Producing kaviar and caviar is an elaborate process, whereas using the whole roe sac is usually simpler, e.g., simply boiled, fried, or prepared as a paste or as a spread.

All kinds of fish eggs can be made into kaviar, whether the roe are from salmon, flatfish, lumpfish, or flying fish. It is the eggs from sturgeon that, internationally and historically, provide the most well-known and sought-after type of kaviar, so-called real kaviar, which in salted form is marketed as—yes—caviar. There are at least 40 different species of fish in addition to sturgeon whose roe are used commercially as kaviar, and there are also roe from other marine animals such as mollusks (e.g., mussels and squid), crustaceans (e.g., shrimp and lobsters), and echinoderms (e.g., sea urchins and starfish). Fish roe have previously been divided into black kaviar to denote real caviar and red kaviar to denote all other types of fish roe, but since caviar can have colors other than black, and other types of fish roe can have a wide spectrum of colors, this is not a good and precise classification. In addition, some fish roe are sometimes dyed black to imitate caviar.

1.6.2 Kaviar Substitutes

Although certain products made from fish roe from non-sturgeon fish are often referred to as caviar substitutes, e.g., roe products as spreads in tubes, we will in the present context by kaviar substitutes mean products made from raw materials that are not fish roe but are nevertheless used in some way to replace kaviar in terms of taste and application. These can be products processed from fish meat, seaweed, certain vegetables, and gelling agents with various taste adjustments. In some cases, the products can resemble fish eggs and in other cases be a paste that is sold as spreads in tubes.

1.6.3 Other Gourmet Roe Products

The most common gourmet roe products in addition to real caviar are, as mentioned above, salted fish eggs released from the roe sac of various fish species. However, there is also a type of very classic gourmet roe product in the form of salted and dried whole roe sacs, which are known from both the Mediterranean area and a number of countries in Asia. This applies, for example, to *bottarga* in Italy (*botargo* in Spain) and *karasumi* in Japan. The roe sacs are taken whole from tuna, swordfish, mullet, or various fish of the cod family, after which they are soaked in sea salt for a few weeks and then hung to air-dry for about a month (cf. Fig. 1.4).

Fig. 1.4 Whole roe sacs from fish hung up to dry into *bottarga*

In some cases, the roe are lightly smoked, marinated, and roasted. Roe products of this type are used in a variety of ways, for example, as a snack, as a side dish, or grated over a dish. It is the umami taste of the roe that comes into play here.

Finally, we also have so-called white kaviar, which are the eggs of land snails. White snail eggs are a great delicacy in many countries, e.g., France and Poland, and they are now also entering the market as a gourmet product in other countries.

Fish Roe as Food During Lent

Fish roe and kaviar from various types of fish were previously in some Catholic countries considered as permitted food during Lent. This is in a way paradoxical, as fish themselves were not always permitted. The opposite situation applies to grapes and olives, which were permitted to eat, whereas the more luxurious and prepared products wine and olive oil were prohibited. But this meant that foods such as caviar, *taramasalata*, and *bottarga* were permitted products during Lent in, e.g., Greece and Crete.

Recipe 1.2 Roe Biscuits (Fig. 1.5)

35–30 pcs

150 g roe
120 g sticky rice, risotto rice, or porridge rice
5 dl water

Fig. 1.5 Roe biscuits

½ tsp salt
2 tbsp yeast flakes (optional)
1 tbsp onion powder
1 pinch cayenne pepper
1 l neutral oil

1. Rinse the rice until it is not milky, place it in a small saucepan, add water and let the rice simmer over low heat under a lid for about 25 min; stir occasionally until there is no more water left.
2. Let the rice stand under a lid for 20 min, place it in a bowl, and stir with the roe, salt, yeast flakes, cayenne pepper, and onion powder.
3. Cool the rice.
4. Roll out the dough approx. ½ cm thick on a baking sheet with baking paper both below and above (this makes it easier to roll out the dough uniformly),

remove the top layer of baking paper, and bake the dough for about 40 min at 120 °C.

5. Remove the sheet and turn the baked rice roe out onto a new piece of baking paper.
6. Cut the dough into small or larger biscuits depending on the use and bake them for a further 30 min at 120 °C, until the dough is dry but still has some moisture in it.
7. Store the biscuits in an airtight container until they are ready to use.
8. Heat oil to 190 °C in a small saucepan and fry a few biscuits at a time for about 10 sec, until they puff up, and place them on absorbent paper.

Roe biscuits are a good way to start using roe that is otherwise unused. All types of roe, especially roe with small eggs, can be used. The crackers can be eaten as a snack or with a dip of, for example, hummus or *taramasalata*.

Recipe 1.3 Cucumber Sandwich with Roe and Cream Cheese (Fig. 1.6)
12 small triangle sandwiches

50 g roe as caviar, e.g. *lögrom/sikrom,* or roe from trout, lumpfish, salmon, or herring
6 slices of spongy toast
2 firm, thin cucumbers without too many seeds
225 g creamy cream cheese
1 lemon
Freshly ground pepper and salt
Fresh dill

1. Slice the cucumbers into thin slices on a mandolin the same length as a sandwich bread and place the slices on a clean cloth to drain the juice.
2. Stir the cream cheese with freshly ground pepper and a little salt and grate some lemon zest into it.
3. Spread the six slices of sandwich bread with ⅔ of the cream cheese and place half of the cucumber slices on top of three of the slices.
4. Place the roe on the last three slices of bread and press the roe into the cream cheese.
5. Place the slices of bread with roe on top of the slices of bread with cucumber.
6. Spread the remaining cream cheese on top and place the remaining cucumber slices on top of the cream cheese, overlapping until they are firm to cut.
7. Wrap the sandwiches tightly in cling film and refrigerate under light pressure.

8. Cut the crusts off the sandwiches and cut each sandwich diagonally into four triangular pieces.

Cucumber sandwiches can be served as an appetizer with a glass of sparkling wine, with a cup of tea, or as a small starter. Feel free to use different types of roe with different colours and textures. You can also use prepared roe from, for example, steamed flatfish or canned cod roe.

Fig. 1.6 Cucumber sandwich with roe and cream cheese

Chapter 2
Eggs and Roe

2.1 Eggs, Kaviar, Gonads, Coral, and Sperm

Fish and many other marine animals reproduce by eggs. The eggs are formed in the ovaries, (cf. Fig. 2.1), and in addition to the germ of a new individual, an egg contains the necessary nutrients in the form of yolk and white for the development of the fetus. The yolk is in principle a single cell, where the germ is the cell nucleus, which contains the genetic material.

In contrast to a single egg, roe refer to collections of immature and mature eggs in the ovaries. Such a collection can consist of a very large number of eggs. The roe are enclosed by an egg or roe sac. In the intact roe sac, the eggs, also called the roe grains, are held together by a thin solution of proteins that act as a kind of connective tissue. The entire sac is surrounded by a thin and fragile membrane. In fish, the roe sacs are shaped like a pair of trousers, with the two "trouser legs" extending backward into the abdomen on each side and the fish's anus exiting through the "crotch" of the trousers. The roe sacs in fish are therefore sometimes called "trousers" (cf. Fig. 2.2). The roe sacs can take up a large part of the mature fish's weight, typically 10–30%.

For some species, e.g., sea urchins, roe refer to the entire reproductive organs (gonads), and in some cases, for food purposes, there is no distinction between the female and male gonads, which can be difficult to tell apart for the lay person. Since the word "gonads" is not very gastronomically appealing, terms such as coral or simply roe and eggs are used, even though the gonads encompass something more. In crustaceans and mussels, the immature roe are almost always referred to as coral, because it can have a reddish or yellowish color. When crustaceans such as lobsters, Norway lobsters, and crayfish keep their eggs for maturing between the small swimlets under their tail, it is said that "they have berries."

O. G. Mouritsen, K. Styrbæk, *Roe and Roe Gastronomy*,
https://doi.org/10.1007/978-3-032-13142-3_2

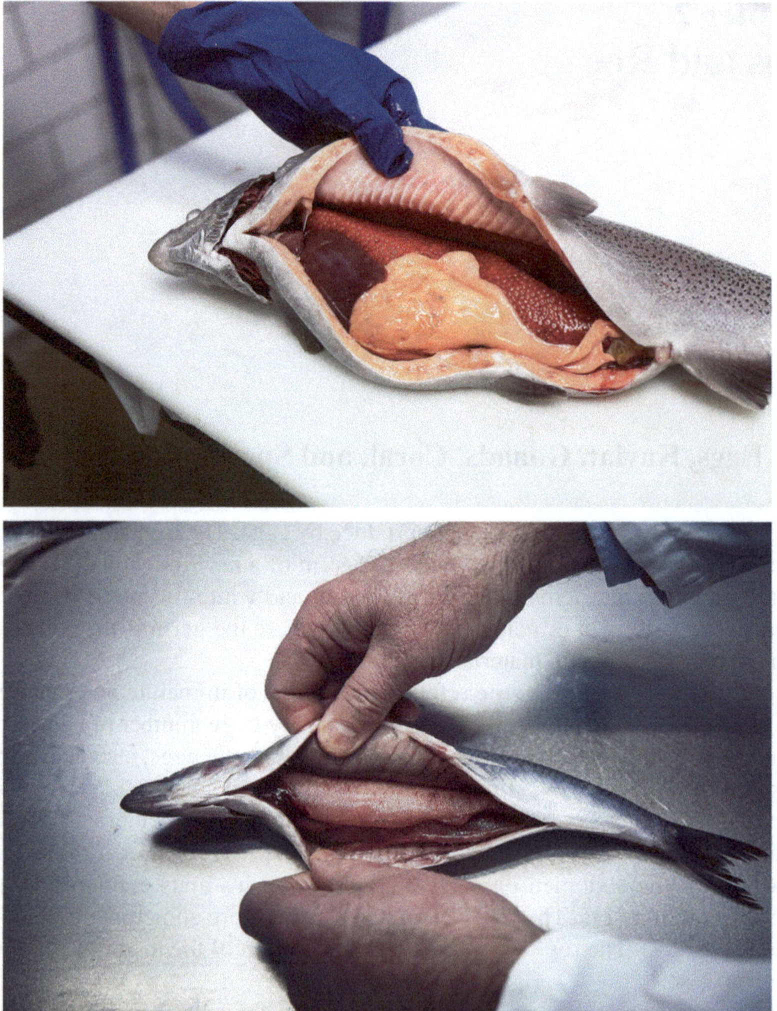

Fig. 2.1 Rainbow trout (top) and herring (bottom), with the abdomen opened and the roe sacs exposed

In some languages, the sperm of the male fish, the so-called milt, is called white or soft roe. In contrast, the eggs are described as firm and hard roe. The sperm mass consists of microscopic sperm cells, and the texture is therefore creamier in contrast to the more or less grainy structure of roe.

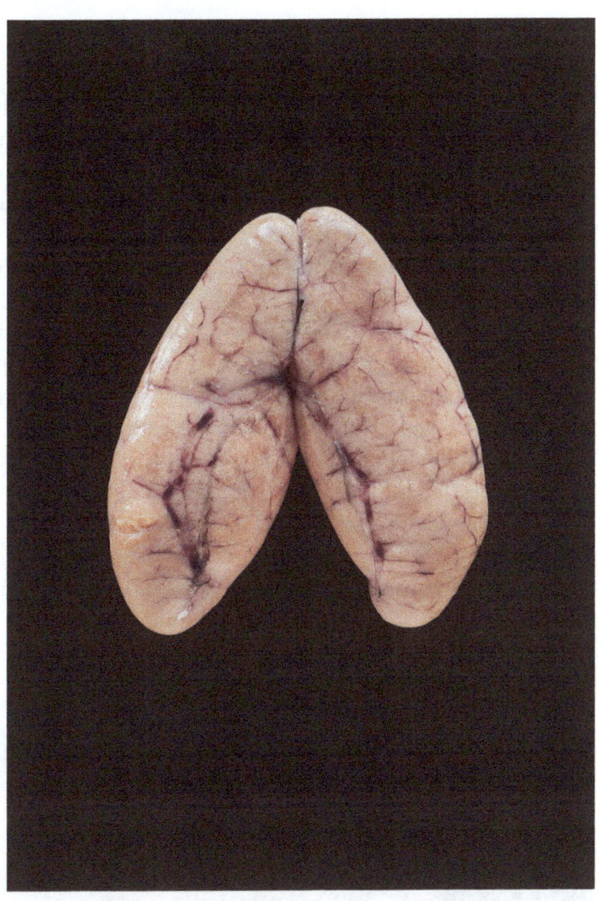

Fig. 2.2 Fresh roe sacs ("trousers") from cod

All Kinds of Eggs Are Eaten

In addition to eggs from marine animals as discussed in this book and eggs from birds such as chickens and ducks, food cultures around the world have eaten eggs from many other kinds of animals. This applies to eggs from reptiles such as turtles and eggs from insects like ants. An example is one of the largest ants in North America, *Liometopum apiculatum*, whose eggs are considered a delicacy in central Mexico and have been so since pre-Columbian times. The eggs are called *escamoles* in Spanish and are also referred to as Mexican kaviar, which in addition to the eggs can also include larvae and pupae. However, the locals use another term for Mexican kaviar, namely, *la hormiga pedorra*, which means something like farting ants. The name is supposed to refer to the sulfurous smell that evaporates from the ants' nest. The eggs are characterized as creamy, buttery, and nutty. Another curious example is the eggs and gonads of jellyfish.

2.2 The Structure of an Egg

Fish eggs are basically structured like bird eggs but without a calcareous shell (cf. Fig. 2.3). Eggs from different fish species have an overall similar composition, consisting mainly of water, protein, and fat (oil), and although there are variations in composition, such as fat content, it is mostly the size of the egg that varies (from 0.25 to 10 mm in diameter).

The chemical composition of fish eggs is typically 70% water, 25% protein, and around 1–2% minerals. The fat (oil) content varies from 5% to 20% and broadly reflects whether the eggs are from fatty or lean fish, and the energy content varies accordingly. There are some types of roe that have a particularly high fat content and a correspondingly lower water content. This applies, for example, to salmon and sturgeon, whose fat content can be as high as 20%, and the cholesterol content is also high. The fat content is important for both the taste and texture of all types of roe products. The texture is particularly determined by the fact that the fats in fish eggs have low melting points and that the contents of the eggs therefore feel very liquid.

Eggs from crustaceans and cephalopods generally have a composition similar to that of lean fish, while eggs from echinoderms such as sea urchin are more similar to fatty fish. The roe from mussels contain much more water and significantly less

Fig. 2.3 Close-up of salmon eggs

protein and fat than roe from fish. Land snails are in a class of their own with very low protein and fat contents in their eggs but on the other hand with high amounts of carbohydrates.

The special thing about the roe from all marine animals is that the fats in the roe have a large predominance of unsaturated fatty acids, especially the so-called super-unsaturated ω-3 and ω-6 fatty acids. It is the ratio, ω-3/ω-6, between the two types of fats that is important for the health value of roe as a food and not just the amount of ω-3 fat. The ratio between the content of the two types of fats in roe from marine animals is around ω-3/ω-6 = 5 but is in some cases much higher. This means that there can be a large overweight of ω-3 fats in roe. This can be compared to the very low ratio ω-3/ω-6 = 0.13 for chicken eggs, which also typically contain three times more cholesterol than fish eggs.

An egg is in principle a single cell that contains the DNA of the genetic material. When the immature egg is in the ovary, it is surrounded by a layer of specialized support cells (a follicle), which provide the egg with nutrition and help in its maturation. During maturation, the yolk grows, and toward the end, it becomes pigmented, and the egg is detached from the layer of follicle cells and finally from the egg sac itself (cf. Fig. 2.4). When the egg is released, a jellylike shell forms around the egg in contact with water, and the egg hardens.

In order for eggs to become kaviar, they must be loose and therefore freed from the ovary, the egg sac, and other membranes. Different caviar/kaviar products use eggs at different stages of maturation. For the production of caviar from sturgeon, the eggs are still surrounded by the layer of follicle cells but close to maturity. The eggs are detached from each other using mechanical techniques, where the membranes are also separated out by sieving. The layer of follicle cells determines the texture of the caviar. For some trout kaviar products, the same is done, and the kaviar comes from relatively soft eggs. For other types of trout kaviar, the eggs are harvested when they are mature, and at this stage they are harder.

The individual egg is made up of an outer membrane and wall of collagen (connective tissue) (cf. Figs. 2.5 and 2.6). The wall is called the zona radiata (or chorion), which forms a rigid network that helps give the egg its shape and appropriate elasticity. The wall is pierced by some pores and in addition has a special, narrow

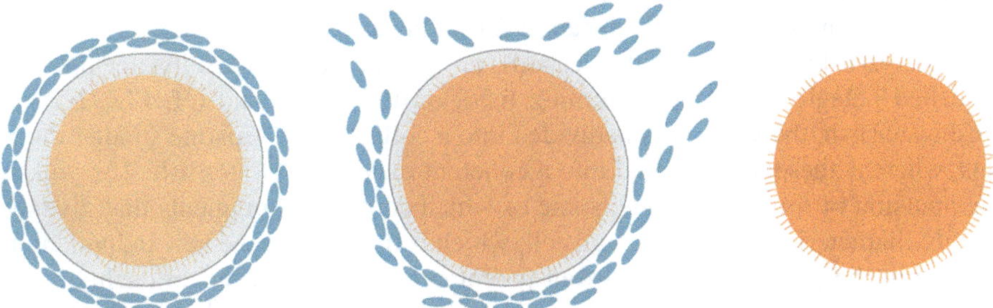

Fig. 2.4 Maturation of fish eggs (from left to right). The immature egg on the left is surrounded by a layer of follicle cells that are released during maturation

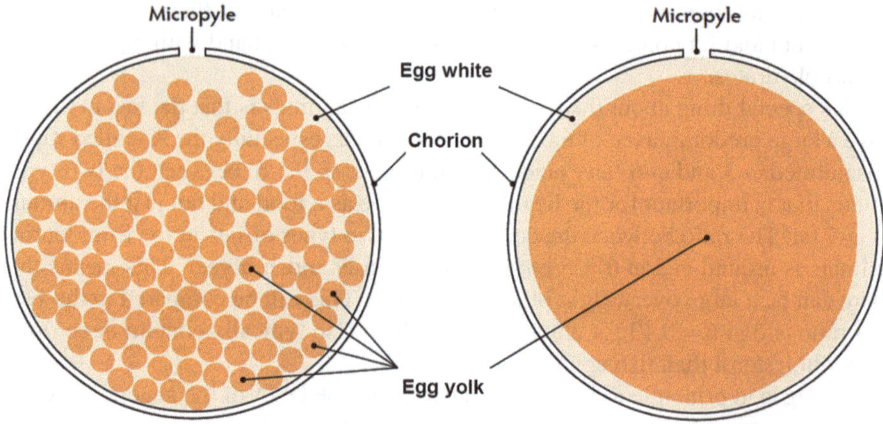

Fig. 2.5 Schematic illustration of fish eggs. On the left, an egg from freshwater fish, where the yolk is divided into separate smaller droplets. On the right, an egg from a saltwater fish, where the yolk is a compact body. The different parts of the eggs are explained in the text

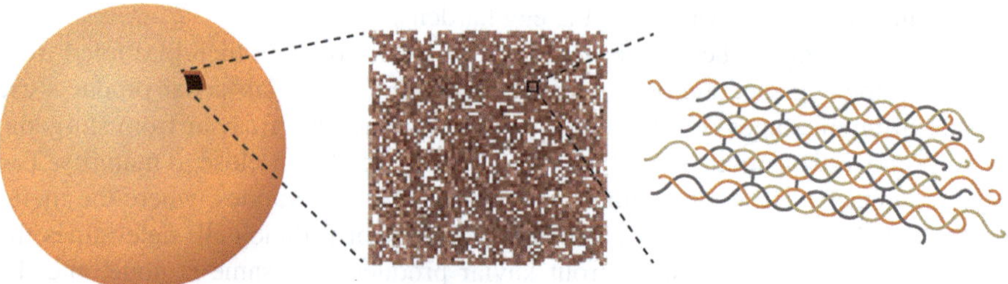

Fig. 2.6 The collagen network (chorion) on a fish egg

channel (micropyle) through which the sperm cells can penetrate during fertilization. Inside the wall lies the egg white, which is an aqueous solution of proteins (ovoglobulin and albumin) that lie outside the yolk itself. The white also contains the substance lysozyme, which has an antibacterial effect. The yolk is separated from the white by a membrane (the vitellin membrane).

The structure of the yolk is different for freshwater fish and so-called demersal fish (fish that live mainly on the seabed), on the one hand, and saltwater fish and so-called pelagic fish (free-swimming fish), on the other hand (cf. Fig. 2.5). In freshwater fish, the yolk mass is divided into a collection of separate smaller droplets, whereas these are gathered into a coherent mass in saltwater fish. The yolk is an emulsion of tiny fat droplets, some of which contain the pigments that give the egg its characteristic color. The egg cell, which is the germ of the new individual, is located in the yolk. In large and transparent eggs such as those from salmon and trout, the egg cell can sometimes be seen as a small "eye" inside the egg. In the egg sac, the individual eggs are loosely held together by a protein solution, and the whole is surrounded by a fragile membrane.

Since fish eggs contain largely the same proteins as chicken eggs, they behave in the same way when heated. In the temperature range of 63–65 °C, the proteins coagulate, and the egg mass becomes solid but retains its taste and aroma. This happens when roe are pasteurized to extend their shelf life.

2.3 Many Eggs in Many Sizes

Some fish let their eggs float freely in the water, where they flow together with plankton in the ocean currents and can hatch far from the mother fish. Other fish lay their eggs on the seabed or stick them to rocks and seaweed and possibly guard them until they hatch. After hatching, the new individuals are extremely vulnerable, and the vast majority never grow to be large before they have become prey for other marine animals or birds or perhaps even eaten by their own species. Depending on the biology and habitat of the individual species, a sexually mature fish must therefore produce a large number of eggs in order to propagate the population. The egg number can range from a few tens of thousands to several million eggs.

Fish produce eggs in numbers within an enormous range of variation. The number depends on both the species and the size of the fish. There is a roughly linear relationship between the weight of the fish for the same species and the number of eggs. A flounder can lay approximately 1850 eggs per gram of body weight, which is equivalent to 1.2 million eggs for a large flounder. The flounder's eggs are approximately 1.1 mm in diameter. At the other end of the scale, we find lumpfish with only approximately 38 eggs per gram of body weight, equivalent to approximately 230,000 eggs for a large lumpfish, whose eggs are twice as large (2.2 mm) as those of the flounder. This may be a bit of a contrast to the fact that we think of lumpfish as having many eggs, but they actually have few in relation to their body weight.

Between the two species mentioned, we find cod, which produces approximately 750 eggs per gram of body weight, but this means that a large cod weighing 15 kg has 12 million eggs in its trousers.

Biologists talk about fecundity, which can be translated as spawning ability and relative fertility. It is usually assumed that there is a reciprocal relationship between egg size and fecundity. The relative fecundity is often defined as the number of eggs per gram of body weight. One of the fish that lays the most eggs is the common mola, which is the world's largest bony fish. The mola lays around 300 million eggs. Yet its relative fecundity is less than that of the flounder.

Egg sizes for different fish species vary between 0.3 mm and up to 8–10 mm (cf. Fig. 2.7 and Table 1 at the end of the book), and there is no correlation between the size of the fish and the size of the eggs. On the other hand, egg size and egg number depend on the habitat in which the fish live and in particular whether the fish species in question lays free-swimming eggs (pelagic fish) or lays eggs on the seabed (demersal fish). Eggs are usually larger for fish species that live in less saline water.

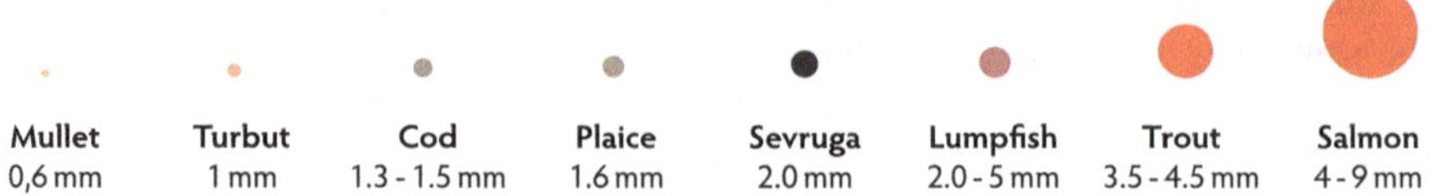

Fig. 2.7 Fish eggs in different sizes

The survival rate of eggs and larvae vary widely in different environments, and the number of eggs and the egg size are adapted so that a species reaches optimal reproductive capacity with the least possible cost. Fish that spawn free-swimming eggs lay far more eggs than fish that spawn on the seabed. This can be explained by the fact that free-swimming eggs are exposed to great challenges in the form of waves and strong ocean currents, the danger of being eaten by other fish, a lower probability of being fertilized, and problems with ending up in environments where the new fish cannot thrive.

The challenges and dangers for eggs and fry on the seabed are less, and fish that spawn there lay fewer eggs, either singly or in smaller clumps where the eggs stick to the bottom and to each other. These fish species have low relative fecundity. Some of these species guard the eggs until they hatch, such as lumpfish, which means that they can manage with low relative fecundity.

In general, the species that lay relatively large eggs also lay the fewest. Among these we find plaice and lumpfish. At the other end are flounder and bream, which lay many but small eggs. In addition, small eggs correspond to high relative fecundity, and large eggs to low relative fecundity. Although the survival rate of larvae from large eggs is greater than that from small eggs, there is no evidence to suggest that there is an evolutionary development toward larger egg sizes. However, in recent years, it has been surmised that the pressure on plaice due to intensive fishing in the North Sea and changing living conditions may be the reason that plaice has increased its egg count. This may indicate that the survival potential of the individual egg is the most important factor for the successful reproduction of the species.

It is interesting to compare two common fish, herring and cod, both of which are very widespread in comparable ecological niches of the sea. Herring lay its eggs on the seabed, the eggs are 1.4 mm in size, and the relative fecundity is around 320. Cod spawns its eggs freely in the water, the egg size (1.5 mm) is comparable to herring, and its relative fecundity is around 750. The two fish species have thus adopted different strategies for reproduction, and cod must have a higher relative fecundity than herring in order to fill its large niche.

A sturgeon can produce from 100,000 to seven million eggs, somewhat depending on the species and especially the size and age of the fish. The relative fecundity is quite high, typically 380 eggs per gram of body weight. The average egg diameter is 2.7 mm with an average weight of 17 mg. Sturgeon generally reach sexual maturity late, and 2–4 years may elapse between each gestation period.

2.4 The Color of Fish Eggs

As with all other raw food items, the first assessment of fish roe and eggs is done visually, focusing on size, color, and optical properties, e.g., clarity, color hue, purity, and shine. When it comes to the color itself, it is especially the degrees of red/green and yellow/blue that are important for eggs. Color and other optical properties can in principle be measured with optical techniques, e.g., spectroscopic, but in practice, at roe producers, this is done manually by people with extensive experience.

Color and optical properties are important quality parameters when selecting and grading fish eggs, and this applies in particular to sturgeon roe and caviar. The values of these parameters depend not only on the fish species but also on the fish's habitat and feed, the age of the fish, the degree of maturity of the eggs, and, not least, the treatment that the eggs may have undergone. Here, salting plays a special role.

2.4.1 Pigmentation of Fish Eggs

Ovaries and roe are almost always pigmented in fish. Exceptions are albinos, whose roe in some cases, e.g., as white caviar from sturgeon, are particularly sought after because it is rare. The lack of pigmentation does not affect the taste and aroma. Pigmentation takes place during the formation of the egg yolk in the ovaries.

Fish eggs occur in a wide spectrum of colors, ranging from yellow, orange, and red to green, blue, and purple. The colors are mainly due to yellow, red, and orange natural pigments from a group of molecules called carotenoids. This is the same class of substances that give color to many vegetables and fruits, e.g., carrots, tomatoes, and corn. In combination with certain proteins, the special colors of the carotenoids are not seen, and blue, green, and purple colors are instead formed. As long as the carotenoids are bound to these proteins, they do not show their typical red, yellow, and orange colors, but if the carotenoids are released and detach from the proteins, e.g., when heated, their colors come forward. This is a phenomenon well known from cooking crustaceans such as lobsters, crayfish, and shrimp. For culinary gourmet products, the roe and eggs are often sorted according to color as shown in Fig. 2.8.

Among the many different carotenoids in fish eggs are astaxanthin, canthaxanthin, zeaxanthin, diatoxanthin, doradexanthin, idoxanthin, tunaxanthin, lutein, and a number of substances derived from astaxanthin. As some of the chemical names suggest, these are inspired by fish names.

Finally, fish roe can contain varying amounts of the substance melanin, which we know as the substance that gives skin and hair dark, brown, and black colors. These dark colors will typically overshadow the carotenoids, and they are crucial for giving sturgeon caviar the characteristic brown and blackish-brown colors.

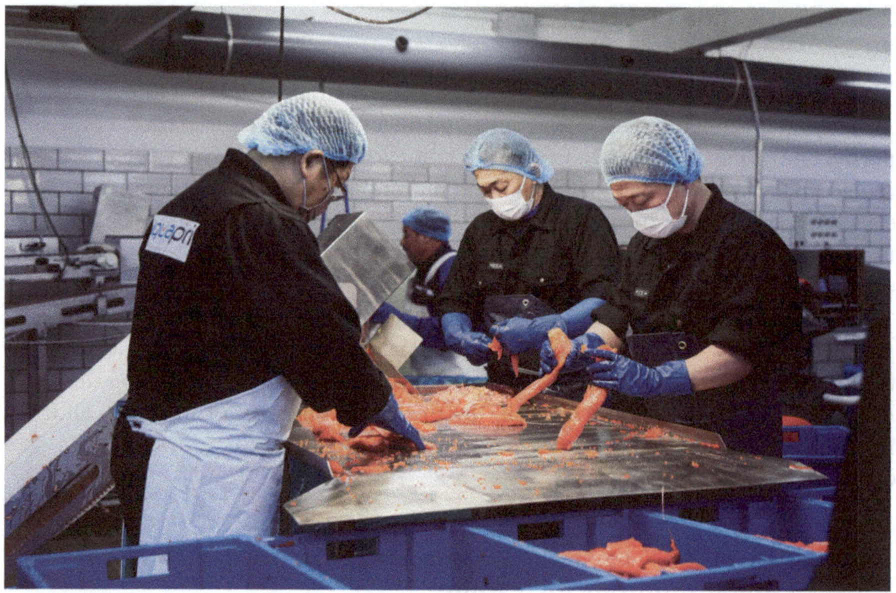

Fig. 2.8 Sorting of trout roe according to egg color

The presence of carotenoids in fish and fish roe is important for several aspects of fish development, including the formation of colors in the skin of the fetus. In addition, carotenoids are important for the hormonal system (the endocrine system) and for the development of reproductive organs (both ovaries and testicles) as well as the maturation, fertilization, and hatching of eggs and their viability. This applies to both fish and shellfish. Some fish species can synthesize their own pigments but not all. For example, species from the salmon family (Salmonidae) depend on obtaining carotenoids from food, especially small crustaceans such as zooplankton.

Carotenoids are fat-soluble, so they are found dissolved in the oil droplets inside the egg yolk. This is useful to know if you want the color from roe to be included in dressings and sauces, where fish eggs are used in emulsions such as mayonnaise.

2.4.2 *Eggs in Many Colors*

The different pigments in roe and the way they work together give rise to the incredible rainbow of colors that the roe and eggs of different fish species have. Information on the colors of the different fish eggs is collected in a table at the back of the book (cf. Table 11 at the end of the book). Here, we will just mention a few examples that illustrate the range of roe colors in different fish species: sardine, yellow; carp, orange, red, green, blue, and violet; flying fish, white yellow; mackerel, yellow brown; Atlantic cod, brown orange; bonito, yellow brown; salmon, red; flounder, red orange; *hamachi*, white yellow; pollock, pink; and sturgeon, from light, golden, yellow, and red colors to brown and jet black. More details on the visual appearance and colors of the different fish roe can be found below for the different fish species (Chap. 4). Discussion of the color of roe and eggs, from sturgeon in particular, can be found in the chapter on caviar (Chap. 3).

Some commercial roe are artificially colored either to imitate caviar or to provide colors that can add a fresh and exciting element to a dish. This can be a brown color with soy sauce or a green color (and strong flavor) from *wasabi*.

2.4.3 When Roe Changes Appearance and Color

Since roe are a food is almost always treated in one way or another, not least with a view to preserving and enhancing the flavor by salting, it is useful to know how the color of fish eggs depends on different preparations and storage.

It is the carotenoids in particular that can be affected by salting and storage, and their color depends on temperature, salt concentration, ripening, and fermentation, because the carotenoids can be oxidized and broken down. These processes can also change the structure of the eggs, which affects their optical properties, e.g., in terms of translucency, shine, and dullness.

If we take salmon roe as an example, the raw and immature eggs are red orange, whereas the mature eggs can vary more toward light orange without changing the degree of yellow. In general, salmon eggs are relatively intense in red color, which is due to a high content of astaxanthin. Due to a relatively low water content, salmon eggs are only slightly translucent compared to other fish eggs, and drying does not change this compared to raw roe.

Salting, drying, and fermentation weaken the intensity of the red color in salmon eggs. The opposite is the case for eggs from, e.g., hake, which are initially whiter and more translucent. However, all roe lose translucency and become duller when salted, which is due to structural changes in the egg.

The degree of red color is the most important factor in distinguishing different roe products, and when it comes to hedonic evaluation of roe, it is the color intensity that has a positive effect. This also means that consumers will prefer the salted and dried products that have lost the least amount of red color and clarity and have retained the intensity of the color. Studies have shown that the color preferences of taste panels, however, do not depend on whether a given color tone of roe is produced by artificial coloring or not.

2.5 Texture, Aroma, and Taste of Fish Roe

Roe and fish eggs are seldom eaten completely raw (so-called green eggs although they are rarely green in color). Before the eggs are ripe, they are small and hard and have only a weak aroma and taste. The texture and taste, and often the aroma, of roe products are therefore typically characterized by whether the roe have been salted, dried, fermented, or aged, processes which lead to a number of physicochemical and enzymatic processes that significantly change the sensory properties of the roe. In addition, the characteristic aroma is often determined by the breakdown products of the eggs' unsaturated fats.

The texture is determined both by the properties of the individual egg and by the fact that they are usually eaten as a lump (a spoonful), i.e., as caviar or kaviar, where there are spaces between the eggs and where the eggs stick to each other with the help of the proteins extracted from the egg white during salting. The overall experience of the texture is therefore quite complex, as described below, and referred to by terms such as creamy, melting, and possibly exploding and crunchy. The roe's own taste first emerges when the eggs burst, and amino acids in particular flow out, giving rise to umami taste but also a touch of slightly sour and bittersweet notes.

2.5.1 Texture and Mouthfeel

As previously described in Sect. 2.2, a fish egg is a kind of container with a firmer, outer membrane and a more liquid interior, which is structured with a yolk membrane (the vitellin membrane) and a yolk with a lot of small oil droplets inside. The texture and mouthfeel will therefore reflect that the egg is a rather complex particle with a certain internal structure, which is not a simple liquid. Furthermore, the liquid (egg white) between the outer membrane and the yolk membrane contains some large molecules, proteins, which regulate how viscous this liquid is. It has a great impact on how it feels to chew the egg and how it feels when the egg explodes and the liquid from both the white and the yolk flows out into the mouth and combines with the saliva. The egg can be hard and brittle, and it can be elastic or yielding and may feel creamy when it breaks. The mouthfeel of the exploded egg is characterized by the fact that the unsaturated fats in the roe have low melting points and therefore feel very fluid.

A physicochemical effect of salting is that salt stimulates a special, naturally occurring enzyme (transglutaminase, which has also been called "meat glue") that works by cross-linking the proteins in the outer membrane of the egg, making it firmer and stiffer and therefore affecting the mouthfeel. Another effect is that the salt weakens the electrical repulsion between the proteins in the egg white, causing them to bind together and thicken the egg white. This thickening gives a creamier mouthfeel when the liquid is released after the egg is broken.

One would think that salting makes the eggs flabbier because salt creates an osmotic pressure that would draw liquid out of the egg in the same way that salting meat and vegetables draws water out of the cells. However, fish eggs have two membranes, an outer one, which is reinforced by a collagen network, and an inner one, which confines the yolk. During salting, a salty liquid is formed between the two membranes, and it helps to "pump" the egg up and make it firmer and more spherical. This is why salted fish eggs are often more spherical than raw eggs. However, there is a complex dependence on the amount of salt, as the egg's liquid content increases at low salt concentrations and then decreases again as the salt concentration inside the egg increases. For example, in Alaskan pollock roe, it has been found that the best sensory effect is achieved at 3.5% salt.

Several Types of Salt: A Stronger Eggshell
The strength of the egg's outer shell determines the texture and how it feels when the shell breaks. Salting is important for this. Since salting usually involves common table salt, sodium chloride (NaCl), there has in recent years been a focus on reducing the salt content in food for public health reasons, as NaCl can cause an increase in blood pressure. Studies have shown that it is possible to prepare Alaska pollock roe with smaller amounts of NaCl by using another salt, calcium chloride, $CaCl_2$. The calcium ions stabilize a calcium-dependent enzyme, transglutaminase, which at low salt concentrations (0.5–3%) enhances the cross-linking of proteins in the shell. The natural content of calcium ions in the egg is not large enough to provide this strengthening of the shell. By adding small amounts of calcium salt, a desired texture can be achieved with smaller amounts of NaCl.

Interestingly, it was not until 2020 that actual quantitative and scientific studies were carried out on the texture of fish eggs, e.g., how caviar feels between the teeth and in the mouth. The German physicist Thomas Vilgis conducted a mechanical experiment on commercial caviar from sturgeon and kaviar from trout and investigated how the eggs deform and eventually burst under the influence of force using a special instrument. He was thus able to relate the melting sensation in the mouth to the plastic deformations and friction between the individual eggs when they are pressed together. In particular, the bursting of the eggs contributes to the special mouthfeel of caviar. At the same time, the bursting releases the characteristic flavors and aromas of the fish eggs as described above.

There is an important difference between the mouthfeel of caviar and the roe of most other fish, which helps to make caviar something special. The individual sturgeon eggs rarely burst with a bang, and the mouthfeel is therefore more velvety and creamier, which supports the melting mouthfeel. The best quality caviar are characterized by the fact that no remaining membrane from the eggs is felt when it melts in the mouth.

Texture Analysis of Fish Eggs: The Science Behind Mouthfeel

Texture analysis is a method that can be used to measure the mechanical properties of eggs and roe. Physicist Thomas Vilgis has performed such an analysis of both sturgeon caviar and trout roe kaviar. The change in thickness of a lump (typically a spoonful) of fish eggs is measured between a fixed plate and a moving plate, which compresses the lump with increasing force (cf. Fig. 2.9). The result of such a measurement is a curve (see the figure) that shows the relationship between the force (external pressure) and the distance between the plates and thus the deformation of the lump of eggs. At first, the individual eggs slide easily between each other, while the force slowly increases. Along the way, some fragile eggs may break. At some point, all the eggs are compressed together into a single layer. After that, the force acts directly on each individual egg, and an increasing force is needed to deform the eggs. When the force has become sufficiently large, the eggs break into pieces. As shown in the figure, this gives rise to three regimes, I–III, each corresponding to the mouthfeel that one would experience from caviar.

I. The lump of eggs deforms and offers a very small resistance, which is due to the friction and adhesion between the individual eggs, and the spaces that were originally in the lump of eggs are gradually filled with eggs. As long as there are still holes to fill between the eggs, the force remains almost constant. The holes are filled with liquid, which contains, among other things, some of the proteins that were originally in the egg white but which have been partially extracted during the salting of the roe. The

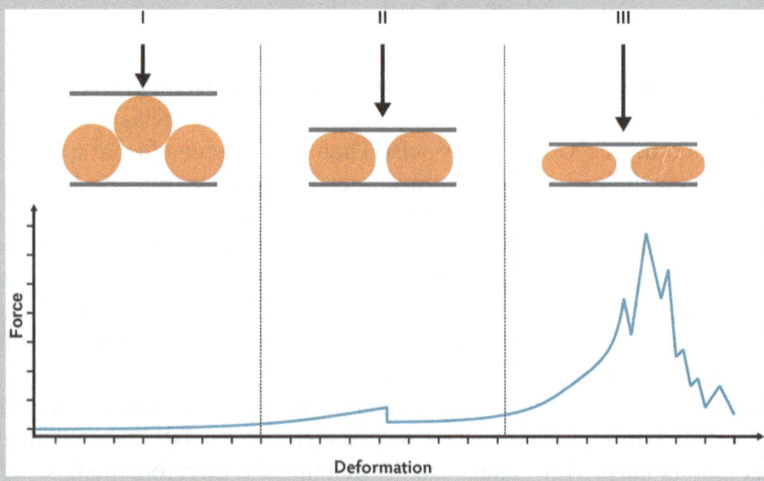

Fig. 2.9 Deformation and explosion of fish eggs

sensory experience of this process is the velvety and "melting" sensation that is felt between the tongue and palate when you have taken caviar into your mouth.

II. There is now only a single layer of eggs left, and it requires more force to squeeze the eggs together. Since the volume of each egg is constant, it is like squeezing a football. It may happen that there are individual eggs that break, either because they are particularly brittle or because they are larger than the rest and therefore more exposed to the force. This can give rise to a small dip in the force curve, as shown in the figure. The sensory experience of this regime is an amplification of the "melting" and creamy mouthfeel, possibly spiced up with explosions of individual eggs.

III. Gradually, a greatly increasing force is required to compress the layer of eggs, and the force curve grows toward a maximum, and more and more eggs break, which gives the jagged shape of the curve. Finally, the eggs no longer have enough strength to be deformed further without breaking. The collagen network shown in Fig. 2.6 is simply torn to pieces. As a result, the force drops sharply, and there are a few jagged edges left on the curve, which signals that even the smallest and strongest eggs must eventually give up. The sensory experience of this process, which is particularly felt between the teeth, is that the eggs exhibit a certain hardness when chewed on and then begin to burst. The serrations on the curve will correspond to the sound of eggs bursting being heard acoustically through the teeth and skull. In the end, only a creamy mouthfeel remains as the oily yolks flow out, possibly interrupted by the collapse of individual recalcitrant eggs.

These three regimes are less evident for caviar than for trout roe, which is due to the stronger adhesion between the individual sturgeon eggs and the greater crowding between the sturgeon eggs due to their more heterogeneous size than trout eggs.

2.5.2 Aroma

As is the case with most other fresh marine products, including fish, shellfish, mollusks, and seaweed, the aroma of completely fresh roe is very weak and subtle, and one is tempted to say that the roe just smells of the sea. Interestingly, the smell or aroma of fresh saltwater or the sea is mainly due to small amounts of decomposition products from small and large algae (seaweed), which, through microbial decomposition, release sulfur compounds that smell pleasant in small quantities. However, in large quantities it smells disgusting (fishy smell, the smell of rotten seaweed).

The well-known and sought-after aroma of caviar and other preserved roe products is primarily due to the breakdown products of proteins and the unsaturated fatty acids in the egg yolk. The aroma is therefore mainly a consequence of preparation and storage. To optimize the quality of the product, a fine balance must be struck for these aroma compounds, some of which are listed in Table 8 at the end of the book.

The nutritional and textural benefits of the egg yolk's high content of polyunsaturated fats must be contrasted with the vulnerability of these fats to being broken down (oxidized) at too high temperatures, when exposed to light and under the influence of oxygen. Although fish eggs contain natural antioxidants, the fats will to some extent be broken down into certain aromas, which are characteristic rancid compounds that are very volatile. This gives aromas that are somewhat nonspecifically described as oily, waxy, floral, and green (like cucumber). Unless antioxidants are added in the form of additives, the best way to delay rancidity is to store roe products at as low a temperature as possible in closed containers before consumption. Some products can be frozen, but the most expensive caviar products will not be subjected to this treatment, so they are stored in special, closed metal cans at around −3 °C until use. The water in the roe does not freeze at this temperature due to the salt content.

In addition, roe also contains some proteins that can be broken down into substances that have flavors of sulfur, overcooked vegetables, stored foods, cheese, and fermentation but can also have a fruity and caramel-like aroma.

Whether you like the term or not, the aroma of the breakdown products of the unsaturated fats and proteins is best described by a kind of faint "fishy smell." An appropriate intensity and character of this fishy smell is simply the aroma of even the finest caviar. Since this type of aroma is found in a wide range of other foods, especially, but not only, marine products such as fish, caviar can be used as an accompaniment to certain other foods, such as chocolate.

The aromatic breakdown products that arise in caviar from the oxidation of the unsaturated fatty acids are commonly found in a very wide range of foods that contain the same fats, such as fatty fish and certain vegetable oils. The same applies to the breakdown products of proteins, especially those with sulfur-containing amino acids. The different sensory expressions of these aromas are described at the back of the book (cf. Table 8 at the end of the book). It has been emphasized that precisely this "common" character of the smell of caviar and other fish roe makes roe suitable in combination with other ingredients in a wide range of dishes in both savory and sweet cuisine.

At least 300 different undesirable volatile aroma compounds have been identified in fish and fish roe, which give what are called "off-flavors." These are odors described as fishy, earthy, woody, musty, tobacco-like, rancid, petrol-like, and rotten (cf. Table 9 at the end of the book). The odors are caused by some strong-smelling substances that one would prefer to avoid in fish products and not least in roe. These substances, and in particular geosmin and 2-methylisoborneol (2-MIB), are produced by some microorganisms (soil bacteria) that are often found in land-based aquaculture installations. The problem is therefore greatest for fish farming in fresh water or in saltwater that is polluted by fresh water. Extremely small amounts of these substances give an unpleasant earthy smell (earthy and musty), which greatly reduces the eating quality of the product. Since the substances are fat-soluble, they accumulate particularly in the fatty tissues and organs of farmed fish, and this also applies to sturgeon, and roe are therefore particularly vulnerable due to its high fat content. The possible accumulation of these substances depends on the fish's feeding status, and when farming fish for roe production, many precautions are therefore taken so that the product does not have the slightest smell of geosmin. However, some would argue that an extremely faint smell of geosmin is characteristic of the aroma of caviar.

Do Not Eat Caviar with a Metal Spoon
Metals such as silver and steel catalyze the oxidation of the unsaturated fats in caviar (and other types of fish roe), and caviar should therefore never be served in metal bowls, and caviar should be eaten with spoons that are not made of metal. Traditionally, special spoons made of horn or bone are used (cf. Fig. 2.10). The salt in the caviar also has a corrosive effect on metal, which can further produce undesirable aromas. For the same reason, special laminated metal tins are used to store caviar so that the caviar does not come into direct contact with metal.

Fig. 2.10 Caviar is best eaten with a horn spoon

2.5.3 *Taste*

The taste impression of roe will almost always be a combination of real taste (described by the five basic tastes on the tongue) and aroma (in the nose). Furthermore, it can be difficult to separate the taste sensation from the mouthfeel described above. The aroma is typically perceived first and is dominant when the eggs have been chewed into pieces. The taste follows afterward, and it is first salt and then a slowly developing umami taste that only slowly fades away.

As mentioned, roe and fish eggs prepared as caviar or kaviar always involve salting. Therefore, the basic taste of salty is the first real taste impression of both caviar and kaviar. Then, when the eggs burst, the umami taste is released. Umami is created by a free amino acid (glutamic acid in the form of free glutamate) and then often in a wonderful synergy with some so-called nucleotides, which we will describe in detail below. Salting stimulates enzyme activity in the raw roe ("the green eggs"), so that free amino acids are produced, which provide taste, not least umami. Knowledge of how much glutamate is in caviar and different types of fish roe are very limited, but it is generally the case that marine fish, shellfish, and mollusks contain significant amounts of free glutamate.

The taste of roe products depends on the process they have undergone in addition to salting. For example, tomato is often added to cod roe, and unnaturally high amounts of glutamate have been found in Japanese cod roe products, suggesting that monosodium glutamate (MSG) has been added.

Umami and Umami Synergy Are Something Special

Umami is, along with sweet, sour, salty, and bitter, a basic taste. We know it best from the taste of, for example, sun-ripened tomatoes, mature and aged cheeses and hams, certain seaweeds, mushrooms, soy sauce, and fish sauce. Seafood in particular can have a lot of umami taste, and this also applies to roe. The special thing about umami is that the taste intensity can increase many times over when there are two different types of substances in the food. Roe can contain large amounts of free glutamate, which gives so-called basal umami taste. But it is not the content of glutamate that alone determines the strength of umami. So-called free nucleotides, such as inosinate (IMP), guanylate (GMP), and adenylate (AMP), can synergistically enhance the umami intensity (cf. also Table 4 at the end of the book). We know this from combinations such as egg and bacon, cheese and ham, tomato and minced beef (sauce Bolognese), and vegetables and meat/bones in a soup stock. Umami synergy means, for example, that while some salmon roe may have small amounts of glutamate, the content of different nucleotides can bring the equivalent umami intensity up to that of caviar (cf. Table 4 at the end of the book).

Caviar and Champagne

Pairing caviar with champagne is a classic combination, just like oysters and champagne. While there may be differing opinions about the taste of such pairings, and while all products are the epitome of ultimate luxury, there is a scientific reason why caviar (or oysters) go well with champagne. The reason is to be found in umami synergy. As a double-fermented and often aged beverage that has had a long contact with the yeast, champagne contains a fair amount of free glutamate. This means that after consuming caviar and the umami taste of the roe has faded, you can reawaken this umami taste by taking a sip of champagne. The glutamate in the champagne works synergistically with the nucleotides left in the mouth by the caviar, refreshing the lingering umami aftertaste. Of course, there are many other components to the overall taste experience, not least the marine saltiness of the roe and the tingling bubbles of the champagne—not to mention the mouthfeel.

Caviar and Chocolate: An Unlikely Pair

Chocolate and caviar may seem like a very unusual and unlikely pair (cf. Fig. 2.11). Yet it is probably the combination that attracted the most public interest when the English chef Heston Blumenthal made himself an international advocate for the so-called "flavor-pairing" theory. According to this theory, which is not a real theory because it has no scientific basis, food and drink should be good companions if they share one or more characteristic aromas. In the case of chocolate and caviar, the common aroma is trimethylamine.

Other well-known examples of pairing are jasmine and pork liver, which share the substance indole; salmon and liquorice, which both contain 2-methylfuran-3-thiol; and chocolate and blue cheese, which also share trimethylamine. The "flavor-pairing" theory was made one of the cornerstones

Fig. 2.11 White chocolate cream with caviar

of molecular gastronomy. However, it has never been clear why the fact that there are common aromatic substances in a pair should make this pair particularly unique in terms of taste, and it has in fact been shown that there is no scientific evidence for this. There is also no consensus that chocolate and caviar actually taste good together.

Many chefs and gastronomes have tried to establish systems, models, and theories for pairing food and drink, not least wine and food. However, a large body of the accumulated experience reflects cultural and geographical conditions, traditions, and the possible embedding of the taste description in the common language, and there is no systematic scientific understanding behind it. Newer, more scientific approaches are psychophysical and are based on the experience of harmony, contrasts, and multisensory interaction between the different sensory impressions, which are linked together in the brain as a complex experienced unity.

The "flavor-pairing" theory received its deathblow in 2011, when a "big-data" analysis of 56,498 different recipes for dishes from around the world showed that while in Western food cultures there is some evidence that many popular recipes contain ingredients with overlapping aroma molecules, the opposite is true for East Asian food cultures. In other words, good pairing is flavored by cultural conditions.

2.5.4 Roe as Both a Flavor and Texture Enhancer

As described above, eggs and roe have their own flavor, aroma, and texture, and these sensory properties can be appreciated by eating the roe pure, e.g., in the form of caviar, cod roe, or lumpfish roe. However, the sensory properties of roe can also be transferred to dishes where the roe are an accessory that, in addition to flavor, also adds texture. The umami taste of roe can be used to add umami to green dishes, and many types of eggs can provide an interesting creamy or crunchy texture. As mentioned in Sect. 1.4, these possibilities give roe an interesting role in the green transition.

Taramasalata is a classic example where roe adds both flavor and texture. A special opportunity to use roe to provide texture exists when making emulsified sauces such as hollandaise sauce or mayonnaise. Normally, a little egg yolk is used for this, which acts as an emulsifier. However, like chicken eggs, roe contains special fats and proteins that can act as emulsifiers, thickeners, or stabilizing agents. You can therefore make both hollandaise sauce and mayonnaise by whisking or blending roe with vinegar or lemon juice. For this to work, the eggs in the roe must be broken in a blender to release their contents. Another classic example of this is thickening of sauces and fish soups with sea urchin roe.

2.6 Additives in Roe Products

In addition to salt, a number of substances may be added to roe products in order to promote taste, texture, color, and shelf life (cf. Table 10 at the end of the book). Salt itself acts as a flavor enhancer, and added MSG contributes to the umami taste. Some roe products from lumpfish, capelin, and cod have thickeners added to give a desired texture. Many types of roe are colored with red, black, and green dyes to mask a possible neutral light and dull yellow color, which some consumers do not find attractive. Finally, preservatives and acidity regulators may be added to prevent unwanted microbial growth of molds and bacteria and oxidation of the unsaturated fats in the roe. Together with salt and refrigeration, these substances help to ensure adequate shelf life in storage and retail stores. Freezing the roe to −20 °C is the most effective way to prevent oxidation of the polyunsaturated fats.

2.7 Fertilization and Maturation of Fish Eggs

Mature and fertilized eggs develop over time toward the larval stage in terms of size, color, and chemical composition. As the water content decreases, the concentration of protein, fat, and inorganic mineral content increases. The details of these changes depend on the species in question. Normally, the eggs will grow in size, e.g., for salmon from 2 to 6 mm. The amino acid composition and the ratio between the different forms in which the fats are found change significantly, although the profile of the fatty acid content varies only slightly, e.g., the important ω-3/ω-6 ratio is little changed during maturation. Immature eggs have a high content of polar phospholipids and triglycerides, but these are converted into free fatty acids during maturation. Histamine levels increase, and cholesterol contents may also change, typically decreasing over time. To quantitatively assess these changes, it is necessary to consider the varying water content of the eggs.

When the roe have become too ripe, e.g., cod roe toward the end of the season, the roe are said to have "eyes." The germ in the eggs has grown into a visible fetus, and the individual eggs are detached from each other. This type of ripe roe are also called running roe. At this stage, the roe are less suitable for eating.

Recipe 2.1 Roe Hollandaise (Fig. 2.12)

1 dl frozen and thawed roe*

- 100 g butter
- A little salt
- 1 tbsp white wine vinegar
- 2 tbsp fresh lemon juice
- 1 pinch cayenne pepper

(continued)

Fresh, finely chopped dill

1. Melt butter in a small saucepan or microwave oven until it is approx. 70 °C.
2. Blend the roe into a fine mass with a hand blender in a cylindrical container together with white wine vinegar, lemon juice, and cayenne pepper.
3. Pour the roe mixture into a round-bottomed saucepan and whisk while the mass heats up over low heat just like for a regular hollandaise sauce.
4. Whisk the roe mass until foamy and hot and then whisk melted butter into the mass a little at a time, first drop by drop and gradually in a gentle stream.
5. Whisk until the sauce sounds hollow and airy, and then season with finely chopped dill and, if necessary, a little extra lemon, salt and pepper.

This recipe uses the emulsifying properties of roe, just like the yolk of a chicken egg. It is therefore important that the fish eggs are blended so they break open. All types of fish roe can be used. Use the sauce for fish dishes, vegetable dishes, or poached roe.

* *The roe are used raw and therefore must be frozen at least 24 hours before use.*

Fig. 2.12 Roe hollandaise with turbot roe on steamed turbot

Recipe 2.2 Toasted Sandwich with Roe, Avocado, and Mozzarella (Fig. 2.13)
Serves 2

4 slices of soft-grain rye bread or rye toast
2 ripe avocados
120 g freshly cooked/poached roe or kaviar such as *löjrom/sikrom* or from salmon, trout, lumpfish, or herring
120 g mozzarella
1 tbsp olive oil
Basil
Freshly ground pepper and salt

1. Cook/poach the roe if using fresh roe.
2. Slice the avocado into thin slices, drizzle with a little lemon, and season with salt and pepper.
3. Place the avocado slices on two slices of bread.
4. Slice the roe and place it and any kaviar on top of the avocado slices.
5. Slice the mozzarella into thin slices, drizzle with olive oil, place the cheese slices on top of the roe, and finish with basil.
6. Place the last slices of bread on top.
7. Put the two sandwiches in a toaster or in the oven for about 8 min at 185 °C.

You can use canned cod roe. This sandwich is a symphony of textures with a crispy crust, creamy avocado, chewy and melting cheese, and creamy or crunchy roe.

Fig. 2.13 Toasted sandwich with avocado, roe, mozzarella, and basil

Recipe 2.3 Eggs with Eggs—A Rustic Roe Meal (Fig. 2.14)
Per person

1 chicken egg
25–30 g or 1 tablespoon of fish roe of your choice
½ spring onion
½ teaspoon soy sauce
Freshly ground pepper, season with a little salt if desired
1 tablespoon oil
Fresh coriander

1. Beat the chicken eggs in a bowl, add soy sauce and pepper, and stir with a fork.
2. Cut the spring onion into fine rings and add them together with 1 tablespoon of roe.
3. Heat a non-stick pan over medium-high heat with a tablespoon of oil, pour in the egg mixture, and let it start to set.

4. Stir slowly in opposite circles with two forks while the omelet sets completely; it can be a rustic and slightly messy dish. It should not be turned like traditional omelets.
5. Let the omelet slide onto a plate, and sprinkle with coriander and, if desired, a little extra roe on top.

The roe has a different texture at the top compared to the bottom, which is heated the most by the pan. This omelette is a real umami bomb with umami from chicken eggs, fish eggs, and soy sauce.

Fig. 2.14 Eggs with eggs—a rustic roe meal

Chapter 3
Genuine Caviar from Sturgeon

3.1 A Short History of Caviar

Caviar based on roe from sturgeon has a rich, stormy, and controversial cultural history. The word caviar comes from the Persian *mahi xâvyâr*, meaning egg-bearing fish (*mahi*), and originally the term *xâvyâr* was used for all kinds of fish roe but gradually became the name for a luxury product from sturgeon. Iranian traders called the product *khavyar* or in Turkish *havyar*, meaning a powerful cake. The modern spelling "caviar" is actually French but is now part of the international vocabulary.

It is rumored that the conversion to Christianity in Russia at the end of the tenth century accelerated the consumption of caviar as a result of the Russian Orthodox Church's injunction of fasting periods, during which roe, but not fish, were permitted to be eaten. However, it was not until Batu Khan, a grandson of Genghis Khan, gained control of the fisheries in the Black Sea and the Caspian Sea in the mid-thirteenth century that caviar was used as a trade name for granulated sturgeon roe.

Two hundred years ago, different species of sturgeon (*Acipenser* spp.) were quite common in the Caspian Sea and the Black Sea and the associated system of large rivers (cf. Fig. 3.1). The same was true of the related species American paddlefish (*Polyodon spathula*) and shovelnose sturgeon (*Scaphirhynchus platorynchus*) in North American rivers and estuaries. Already in the eighteenth century sturgeons were captured in New Jersey and their roe processed into caviar. Caviar, although sought after, was not an uncommon food and could even be found as appetizers at British pubs and in American saloons. However, due to overfishing, damming of rivers, increasing pollution, and poaching, the wild populations declined, and several species became endangered and close to extinction. Already around 1900, caviar started to become a rarer and more expensive product.

During the twentieth century, the market for wild caviar almost collapsed, and the production of caviar from wild sturgeon declined by 90% in the period 1984 to

O. G. Mouritsen, K. Styrbæk, *Roe and Roe Gastronomy*,
https://doi.org/10.1007/978-3-032-13142-3_3

Fig. 3.1 Caviar produced in the late 1900s under primitive conditions from sturgeon caught in the Volga River in the Astrakhan region near the Caspian Sea

1999 which drove up the price of caviar considerably. During the same time, aquaculture of sturgeon picked up. Sturgeon farming for caviar started in the USA in the end of the 1970s, and in the same period aquaculture also increased in Europe.

The market for wild caviar is now heavily regulated, and most caviar on the world market now comes from aquaculture, which experienced a boom from the 1990s onward and made sturgeon production one of the fastest growing aquaculture sectors at the end of the twentieth century. A significant advance was that farmers managed to accelerate the sturgeon's life cycle under controlled conditions in the farms. Today, China is by far the largest producer of caviar followed by Italy, France, Russia, and USA. The major consumers of caviar are USA, Japan, Russia, and China. In Europe, France, Germany, and Spain are leading consumers. The large production has led to a local demand for sturgeon meat but less so in Western Europe.

3.2 True Sturgeons and False Sturgeons (Paddlefish)

Sturgeons are a group of different species that have ancient origins and are therefore sometimes called "living fossils," as they have hardly changed much over many millions of years. Over time, many species have become extinct, and today there are only two families that together include 29 different species. Sturgeon fish are the world's largest freshwater fish; they live a long time and only reach fertile maturity later in their lives.

The class of ray-finned fish (Actinopterygii) arose about 400 million years ago and now makes up about 99% of all living fish with more than 32,000 different species. An order (Acipenseriformes) in this class was formed in the geological era of the Jurassic about 200 million years ago. About 184 million years ago, this order split into three families: Acipenseridae, whose members we now call true sturgeons, and Polyodontidae, which are called paddlefish. The third family is extinct. At the same time as Acipenseriformes, another order (Polypteriformes) was formed, whose members are called bichirs and armored eels, which can therefore be seen as distant relatives of sturgeons but of equally ancient origin.

The sturgeon family Acipenseridae has 27 different living species, whereas there are only two species in Polyodontidae, the American paddlefish (*Polyodon spathula*) and the Chinese paddlefish (*Psephurus gladius*). The Chinese paddlefish has not been observed alive since 2003 and must therefore now be presumed extinct. Paddlefish are called false sturgeons. True sturgeons and paddlefish were some of the species that survived the massive extinction of species on Earth 65.5 million years ago at the transition between the Cretaceous and Tertiary eras, when the dinosaurs disappeared. Sturgeons are thought to have remained virtually unchanged ever since. The sturgeons and the paddlefish are therefore ancient species, and it is believed that their genetic evolution has been very slow over millions of years, and they have very few close relatives.

The very appearance of sturgeons alone suggests that they are of ancient origin (cf. Fig. 3.2). They do not have scaly skin but armor-like bony plates arranged in five rows. They have a tail fin like a shark, and some sturgeons have four barbels to feel for prey on the seabed, where they feed on shellfish and small fish.

Fig. 3.2 The beluga sturgeon (*Huso huso*) is a true sturgeon

Most sturgeons live in freshwater habitats, while others live close to the coast. Many are usually anadromous, meaning they live in river deltas and migrate up rivers to spawn, much like salmon. Sturgeons are the world's largest freshwater fish, and they can also be the slowest growing and longest lived. Some species grow to only half a meter in length, but females of the largest species, the beluga, can grow to 4 meters and weigh more than half a ton.

3.3 Endangered Species

Sturgeons have proven themselves to be survivors over millions of years, and it is against this background that we should be concerned about their present status as some of the world's most endangered animal species. Virtually all wild species of sturgeon are marked in red as endangered species all over the world. Some species are so critically endangered that it is forbidden to catch them or export their roe. This situation is due to the high demand for their roe but also to the fact that wild populations are threatened by shrinking habitats, dams, overfishing, illegal fishing, pollution, and diseases. It is particularly the sturgeon's slow growth and high age of maturity that weaken its adaptability and ability to rebuild declining populations.

As a consumer, you can ensure that the caviar is not illegal or from endangered species by checking if it carries a so-called CITES label (Convention on International Trade in Endangered Species of Wild Fauna and Flora) (cf. Sect. 3.5.7).

3.4 Sturgeons and Their Roe

As mentioned above, there are presently 27 different species of true sturgeon and a single living species of paddlefish (the American paddlefish, *Polyodon spathula*). In addition, there are an unknown number of subspecies and hybrids. The vast majority of sturgeon species are presently considered critically endangered species, and some are on the verge of total extinction. Just under half of the different sturgeon species are presently fished commercially or farmed for caviar production. The large and longest lived species are particularly sought after because they have the most ideal eggs to produce high-quality caviar. Some of the most important and well-known species and their caviar are listed in the following.

3.4.1 The Shovelnose Sturgeon

The shovelnose sturgeon (hackleback, *Scaphirhynchus platorynchus*) is a true sturgeon. It also has the structure of a primitive fish with skin without scales, a tail fin like a shark, and a cartilaginous internal skeleton. In contrast to the sturgeon's barbels, the shovelnose sturgeon has, as its name suggests, a shovel-like nose, which has sensitive electrical receptors to guide the visually impaired shovelnose to food, which it searches for in the muddy bottom. The shovelnose is a filter feeder, meaning it feeds on zooplankton, larvae, small fish, and insects. This characteristic can be an advantage when it comes to providing sustainable feed for aquaculture of this species, unlike other true sturgeons, which can not only live on small creatures in the mud but also eat live small fish.

The shovelnose sturgeon is only found in the USA. It evolved in river deltas and estuaries around the Gulf of Mexico and from there moved up into the major North American rivers, so that wild shovelnose sturgeons are now found in 22 states in the USA and as far north as northern Montana.

The shovelnose sturgeon is a smaller fish than other sturgeons, weighing a few kilos up to a maximum of 5 kg and growing to 20–40 cm in length. It moves up the Missouri and Mississippi rivers and is considered a vulnerable, but not yet endangered, species. It has a typical age of maturity of 8 years but for some up to twice that. It has an average lifespan of just under 50 years, and it spawns every 2–3 years. It is one of the few true sturgeon species from which caviar is commercially harvested from wild stocks. The eggs are black, firm, and quite small, with a mild and nutty flavor.

3.4.2 The American Paddlefish: A False Sturgeon

The American paddlefish (*Polyodon spathula*), also called the Mississippi paddlefish, is not a true sturgeon, although it produces excellent caviar. It is considered the best species of non-sturgeon for producing high-quality caviar, even at a reasonable price. It is found mainly around the Mississippi River and in estuaries out to the Gulf of Mexico, and it thrives in both fresh and brackish water.

The American paddlefish, which is found only in North America, is no longer considered a threatened species but a vulnerable one, and this is due to the authorities' successful attempts to regulate the fishery and make it more sustainable. The distribution of paddlefish was at its lowest in 2004. A major reason for the overexploitation of wild paddlefish stocks is due to pressure from the international caviar market, as American and international regulations began to apply to the production and trade of caviar from true sturgeon.

The American paddlefish typically grows to 1.5 meters in length and weighs about 30 kg when fully grown. It feeds by filtering small animals in water and mud. It typically lives for 30 years but can live up to 60 years and is not sexually mature to produce eggs until the age of 10, which then occurs every 2–3 years. The eggs are harvested from wild stocks. The eggs are softer and smaller than for true caviar, about 2 mm in diameter, and the taste is slightly spicy and nutty with an earthy aftertaste.

3.4.3 Beluga Sturgeon

The beluga sturgeon (*Huso huso*) is a true sturgeon. It is the largest of all sturgeons and can grow up to 5 meters in length and in exceptional cases weigh up to 2 tons. It is the world's largest freshwater fish and the second largest bony fish, and it lives in both rivers and oceans, exclusively as a predator. The beluga is found in the Black Sea, the Caspian Sea, and the Adriatic Sea, but the wild beluga is largely extinct today, and it does not reach sexual maturity until it is about 35 years old. The roe of a sexually mature female can weigh up to 25% of its body weight, and a beluga will typically produce 10–20 kg of roe. Beluga has been successfully bred in river systems, for example, in Bulgaria, and it has been possible to achieve sexual maturity in the 18th to 20th year of life. It can produce roe every 3–4 years. Breeding belugas is not without problems, because, as a predatory fish, the fish eats its own species.

Beluga caviar is the rarest and most expensive of all caviar. The individual eggs are gray or blue black and quite large, with an average diameter of about 3.2 mm. The eggs are fragile, as they have a very thin membrane. The taste is nutty with hints of fruit and flowers. The texture is smooth and silky. Hybrids of beluga, e.g., with Siberian sturgeon, are suitable for aquaculture.

The World's Largest and Oldest Beluga

It is being rumored that the largest beluga ever caught was in Astrakhan on the Volga River. The fish was as long as 12 Russian fishermen standing shoulder to shoulder, and there was said to be 510 kg of caviar in its belly. It is also said that a beluga can live up to 118 years.

3.4.4 White Sturgeon

The white sturgeon (*Acipenser transmontanus*) is North America's largest freshwater fish and is found in estuaries along the northwest coast, from Mexico to Alaska, and up the Columbia and Fraser rivers. It is also farmed in aquaculture, for example, in Idaho. It does not reach sexual maturity until about 10 years old and even later in wild conditions. It can take up to 10 years between each spawning. Its maximum weight is about 800 kg, and it can live to be very old, i.e., over 100 years old. The eggs in its caviar are slightly smaller than those from beluga but have the same color and are oilier and saltier.

3.4.5 Kaluga

The kaluga species (*Huso dauricus*; *Acipenser schrenckii*) are of the same genus as beluga and are also called river beluga, river sturgeon, or Amur sturgeon as they live around the Amur River on the border between Russia and northeastern China. These species move in both fresh and salt waters (semi-anadromous). In aquaculture, they reach sexual maturity after about 7–12 years and even more rarely in the wild. There is a 4–5-year interval between spawning periods.

The kaluga species are almost as large as beluga, and the eggs are comparable but with a more golden color. The egg size is 3.2–3.6 mm. The caviar has a creamy and mild taste. The most sought-after quality of kaluga caviar is from a hybrid, *Acipenser dauricus* × *schrenckii*, which actually produces a caviar that is among the three most sought-after qualities. This sturgeon thrives well in aquaculture with natural mineral and clayey water flow, and the yield can be up to 30 kg of caviar per fish. The eggs, which are quite firm and have a soft buttery taste, can vary in color from black to light golden with brown shades and have a diameter of 3.0–3.2 mm. The light qualities are most sought after.

3.4.6 Osetra

Osetra caviar (oscietra) (Russian sturgeon, Danube sturgeon, *Acipenser gueldenstaedtii*; Persian sturgeon, *Acipenser persicus*) is the common name for caviar from two different species, Russian sturgeon and Persian sturgeon, which were previously thought to be one and the same species because of their similarity. After beluga, osetra is the best-known caviar in the world. Both sturgeon species are found in the Caspian Sea and in Russian rivers. They become sexually mature late, after 9–11 years in aquaculture, and spawn every 4–5 years and are therefore considered less suitable for aquaculture. Osetra can grow up to 2 meters long and typically weigh 80–100 kg but in rare cases almost twice that.

A mature female's weight can consist of up to 10% caviar, but typically 2–4 kg of roe can be harvested from a female. The eggs are 2–3.5 mm in diameter with grayish-black to yellow-brown shades, sometimes with a golden sheen. The taste of the caviar is nutty and buttery and is said to have an oyster scent. Russian sturgeon is today the most sought-after type of osetra in terms of quality and constitutes a large part of the world market. In 2020, the Russian sturgeon was crossed with the American paddlefish to the so-called sturddlefish, which may prove suitable for sustainable aquaculture of caviar.

3.4.7 Sevruga

The sevruga sturgeon (starry sturgeon, *Acipenser stellatus*) is found in the Black Sea and the Caspian Sea. It is somewhat smaller than the Russian and Persian sturgeon, growing up to 2 meters in length and typically weighing 80 kg. It reaches sexual maturity early, after 4–5 years, and is therefore suitable for aquaculture. Each female typically produces 2–3 kg of caviar, and up to 10% of the mature female's weight can be made up of roe. The eggs are dark and grayish black and quite small, about 2 mm in diameter. The caviar has a less complex flavor than osetra. Sevruga caviar is the third of the classic Russian types of caviar, after beluga and osetra, and is usually saltier and produced in the largest quantity of the three.

3.4.8 Siberian Sturgeon

The Siberian sturgeon (*Acipenser baerii*) is naturally found in rivers in Siberia. It is slightly smaller in size than the Russian and Persian sturgeon but matures faster, is more robust, and requires less space, making it suitable for aquaculture. It can live up to about 60 years. France is the largest producer of caviar from Siberian sturgeon. The eggs are firm and medium sized with a brown to black color (cf. Fig. 3.3). The taste is nutty with a hint of *shiitake*. Caviar from Siberian sturgeon today makes up the largest part of the world caviar market, partly because in aquaculture it can be pressed to sexual maturity quickly after only 6–8 years compared to 19–20 years in the wild.

Fig. 3.3 Caviar from Siberian sturgeon (*Acipenser baerii*)

3.4.9 Sterlet

The sterlet (*Acipenser ruthenus*) is a small sturgeon species, weighing no more than 16 kg and reaching a maximum size of 1 meter. It has the shortest lifespan of all true sturgeons, up to about 20 years. Wild populations of sterlet are found in the rivers of Eastern Europe and Siberia. This small sturgeon species is raised in fish farms and becomes sexually mature after 4–5 years and spawns every few years. The eggs are very small, about 1 mm in diameter, light to dark gray in color with a slight silver sheen. The caviar is described as mild in fat and oily with a velvety texture.

An Unusual Cross Between Sturgeon and Paddlefish: Sturddlefish

It is not unusual for closely related fish species to cross (hybridize) and produce viable offspring, and there are also a number of hybrids between different sturgeons in the same family (Acipenseridae), for example, between a female beluga (*Huso huso*) and another male sturgeon, for example, Siberian sturgeon (*Acipenser baerii*) or sevruga (*Acipenser stellatus*), both species that can produce caviar. In fact, sturgeon hybrids account for 35% of global sturgeon meat production and 20% of caviar production.

But it was a big surprise when a Hungarian American research team reported in 2020 that they had succeeded in crossing two species from two different families, Acipenseridae and Polyodontidae, which therefore have their last common ancestor 184 million years ago. The two species are the Russian sturgeon (*Acipenser gueldenstaedtii*) and the American paddlefish (*Polyodon spathula*). The reason why this turned out to be possible is probably that a so-called genome duplication has already occurred in the common ancestor of the two species and that the evolution of the two families has been extremely slow. The crosses have both triple and quintuple chromosome counts in the genome.

The result of the cross has been somewhat amusingly called sturddlefish, and the appearance of the new hybrid is indeed reminiscent of a smaller sturgeon with an upturned shovel-shaped "beak," which is smaller and wider than that of the paddlefish. It is not yet known whether the cross can become fertile or is sterile, as is often the case with hybrids. It is also unclear whether sturddlefish will be suitable for aquaculture. The perspective of this new sturddlefish is that it exhibits characteristics that are a mixture of its parents not only in appearance but also in terms of eating behavior, so that to a certain extent it has inherited the paddlefish' ability to eat plankton and therefore does not need to be fed larger animals such as leeches, worms, mussels, crabs, and small fish. If sturddlefish can produce caviar in aquaculture, the product will therefore be more sustainable than traditional caviar from sturgeon.

3.5 Caviar

Caviar as a luxury product is believed to have first been produced in Russia around 1200, and Russian caviar from the rivers that flow into the Caspian Sea is still one of the most sought-after today. Here, wild stocks of a small number of different sturgeon species were caught by Russian and Iranian fishermen, who then harvested the roe and processed it into caviar, originally presumably as a way of preserving the roe with salt. The salting also improves the taste and texture. The salt content was therefore high, typically 10–20%, and the caviar was packaged and traded in wooden barrels or boxes.

The modern caviar industry is associated with the Russian Empire, and the tsars' monopolization of caviar throughout the seventeenth to nineteenth centuries still stands as a trademark for caviar and the extravagance that surrounds caviar as a luxurious delicacy.

As mentioned earlier, until about 200 years ago, sturgeon was a common fish in the large rivers of the Northern Hemisphere, and caviar was less unusual and much cheaper than it is today. In the nineteenth century, caviar was so common and cheap that American saloons used caviar to stimulate beer drinking in the same way that salted nuts are used today.

3.5.1 *From Sturgeon Roe to Caviar*

There are three ways to extract the roe from sturgeon for the production of caviar. The classic method consists of removing the entire ovary from a slaughtered fish (Fig. 3.4). In modern facilities, the sturgeon is first cooled in water to slow its movements, so that the fish approaches a state of unconsciousness. It is also used to stunning the fish with electric shock. It is then slaughtered and cleaned in clean water. After an incision in the abdomen, the roe sacs are removed within a few minutes. There are usually two roe sacs which are cleaned, and the meat, bones, and skin of the fish can then be used for other purposes.

Newer methods, so-called humane slaughtering methods, avoid slaughtering the fish either by removing the roe with a form of caesarean section or by stroking the roe out of the fish after possible treatment with hormones. In the caesarean section method, a small incision is made in the abdomen of the female fish, and the eggs are carefully removed. The incision is sewn back together, and the fish can live on if it is not exposed to serious infections. There are examples of this being done a number of times on the same fish, but generally you cannot always count on the fish bearing eggs again after a caesarean section.

In the stroking method, the eggs are massaged out of the fish to simulate natural spawning. A prior hormone treatment can ensure that the eggs are ready to be released from the ovaries. Immediately after being harvested, the eggs must be cleaned in a calcium solution so that their membrane becomes strong enough to

Fig. 3.4 Exposing the roe of a sturgeon

withstand the subsequent treatment with salt. Unfortunately, this can sometimes cause an undesirable off-flavor to the product, and the eggs can end up being too hard, which changes the mouthfeel of the caviar.

The classic method is considered to give the best quality of the final caviar product, and most sturgeons that provide roe for caviar production must therefore let their lives be taken away by having the entire ovary removed at slaughter. The loose, "green eggs" are now released from the sturgeon's roe sacs. This is done by pressing the roe through a net or sieve into a container, thereby removing the membrane that holds the eggs together (cf. Fig. 3.5). Both the net and container are made of stainless steel to prevent oxidation of the eggs. Previously, nets made of cotton or nylon thread were used. The eggs are then finely sorted into different qualities, any impurities are removed, and overripe and damaged eggs are removed and used for other caviar products, e.g., pressed caviar as described below. In contrast to this form of pressed caviar, loose eggs are called grain caviar.

The best roe for subsequent salting is that in which the roe are almost ripe for natural spawning. If the eggs are too ripe, they break easily, and the fats are more oxidized and the taste is correspondingly more "fishy." If the eggs are too far from ripe, they are smaller in size, harder, and have little aroma.

The sorted eggs are now quickly washed in a cold 3% salt solution to remove impurities and damaged eggs. The eggs are then dry salted by turning the salt by hand for a few minutes (cf. Fig. 3.6), so that a final concentration of between 3% and 10% is achieved. Unsalted roe will go bad very quickly. The salted eggs are spread out on a fine-mesh stainless-steel mesh, and the eggs are left to drain for 5–15 minutes. The finished caviar is then packed tightly in special laminated metal cans. From the time the roe are removed, it usually takes less than two and a half

Fig. 3.5 The eggs from sturgeon roe are released by pressing the roe through a special sieve

Fig. 3.6 Salting sturgeon roe by hand

hours for the product to be packaged. The cans are refrigerated at −2 °C to −3 °C until use. Due to the salt content, the caviar does not freeze at these temperatures. The cans of caviar are stored for at least 3 months so that the eggs can fully absorb the salt and release the desired flavors and aromas. In some cases, the can is left unsealed for up to a month to achieve the desired degree of oxidized aroma. The salting is absolutely crucial for the taste, texture, and preservation of the caviar, and the less salt, the shorter the shelf life. Too much salt disguises the roe's own flavor. The finest and most sought-after caviar is therefore one that is mildly salted.

During salting, it was previously common to add small amounts of boric acid or its sodium salt, $Na_2[B_4O_5(OH)_4]\cdot H_2O$, which is called borax. Borax acts as a preservative and also gives the caviar a little more sweetness and a fresh taste. Several countries, such as the USA, have banned the use of borax in caviar because borax is toxic in large quantities. If borax is added, the caviar cannot be sold with an organic certificate.

Caviar: But What About the Rest of the Fish?
Already in the Middle Ages, caviar and sturgeon meat were sought-after products, and the English King Edward II (1285–1327) declared the sturgeon a royal fish. To this day, all sturgeons in English rivers are considered the property of the crown. In Europe today, it is mainly in Bulgaria, Ukraine, Serbia, and Romania where sturgeon meat is valued. In addition, Russians and Chinese have a centuries-old tradition of eating sturgeon. In modern caviar production, farms only become profitable if the rest of the fish is also used. This applies to the whole fish for the males after they have been sexed and parts of the fish when the females have been slaughtered for caviar production. For every kilo of caviar produced, 20 kilos of meat is produced. In addition to meat for human consumption, extracts for the cosmetics industry, gelatine, medical products, and leather are also produced from the carcass. With careful planning, all parts of the fish can be used, thus ensuring better sustainability.

3.5.2 Salting of Caviar

There are three methods for salting roe, basically corresponding to increasing amounts of salt and correspondingly longer shelf life but also deterioration in eating quality.

The first method, called *malossol*, which is a Russian term meaning "lightly salted," is associated with the finest caviar, which has only 2.5–3% salt. *Malossol* is considered the best method for preparing and partially preserving roe, so that the subtle flavors and aromas of the roe are optimally maintained without compromising shelf life. But it is a delicate balance. Most caviar, as well as other kaviar products made from fish eggs, are now produced using the *malossol* method. To increase the shelf life of this lightly salted caviar, preservatives are sometimes added to extend its shelf life.

Another method leads to pressed caviar (Payusnaya), which contains up to 7% salt and is a kind of jam-like, spreadable paste with a stronger taste and aroma than caviar, which is mainly due to overripe and damaged eggs with oxidized fats. One third of the eggs in Payusnaya are often broken. During the pressing, a large part of the salt, liquid, and oil is lost, so that only about a quarter of the original weight remains. This intensifies the taste and aroma considerably, often to such an extent that many people do not like the product, but it is still preferred by some connoisseurs.

A third method uses even more salt, up to 8%, and is called semi-preserved or salted caviar but is mostly used for roe from fish other than sturgeon.

Some caviar is also produced and preserved with both salt and a preservative (sodium benzoate). Some caviar is pasteurized in tins in hot water at 65–70 °C to extend its shelf life, but the heating causes the texture to become tougher and the aroma to be less delicate. At temperatures above 70 °C, the proteins in the eggs denature, and the texture of the caviar becomes ruined and rubberier. Finally, there is a low-quality product, Jastichnaja, which is made from eggs that are either far from ripening or overripe or that are poorly separated from the ovaries and therefore have more connective tissue.

3.5.3 Tinned Caviar

Consumers usually encounter caviar in metal tins that are vacuum-sealed. If the tins are kept at around −2 °C, the caviar can be kept for up to 3 months without any changes in its organoleptic properties. Most households cannot refrigerate to such low temperatures, and the general advice is therefore to place the purchased tins in the coldest part of the refrigerator and eat the caviar within a month.

Before serving, it is a good idea to leave the unopened tin at room temperature for 10 minutes, after which it is opened and the caviar can be served in the tin, possibly on ice. An opened tin may be wrapped in plastic film and stored in the refrigerator for up to 5 days.

Caviar producers use significantly larger tins for storing and transporting caviar. These tins contain 1 kg of caviar and are sealed using a special method, which is an old Russian invention. The sealing is carried out using a physicochemical technique in the following way. The tins are cylindrical and filled with caviar with a top, so that there is a little more than 1 kg of weight (cf. Fig. 3.7). The lid consists of another cylindrical tin with a slightly larger diameter, so that it can be pushed down around the tin with caviar like a tight-fitting collar. The very narrow space between the two tins now acts as a capillary space that automatically sucks excess liquid out of the tin's contents and therefore presses the lid down and compresses the caviar, so that the entire tin is sealed. This reduces the access of oxygen to the tin's contents, and the final caviar weight of the tin now corresponds to 1 kg. The large tins can be stored in the refrigerator for up to 6 months, and the content can be divided into smaller tins as mentioned above. The caviar in the large tins, which are closed by this technique, matures very slowly, and it is a matter of consumer preference when the product is considered to have the best taste quality.

Fig. 3.7 Original 1 kg large caviar tin with top of caviar before closing

On a Field Trip: French Caviar from Farmed Sturgeon

France is today the world's third largest producer of caviar from Siberian sturgeon (*Acipenser baerii*), and the annual production for the international market is around 27 tons (2022). In addition, around 8 tons of osetra caviar is produced from Russian sturgeon (*Acipenser gueldenstaedtii*). The sturgeon is raised in aquaculture, and the company Groupe KAVIAR was a pioneer when it started farming more than 30 years ago. The company now has nine different fish farms, hatcheries, and caviar production around Saint Fort sur Gironde, north of Bordeaux in the Aquitaine region. The caviar is sold under the brand name Sturia. Other sturgeon farms in the area account for a smaller production.

Groupe KAVIAR has succeeded in establishing a major European production of caviar from Siberian (*Acipenser baerii*) and Russian sturgeon (*Acipenser gueldenstaedtii*) (Fig. 3.8). The company produces around 18 tons of caviar (2022) annually, of which 11 tons is from Siberian sturgeon. The high quality of the caviar is due to the company's strong focus on research-based and technological development of feed, farming facilities, processing of the roe into caviar, and quality control. The company also has a small production of beluga caviar based on fish that are 15–20 years old.

Fig. 3.8 A Russian sturgeon (*Acipenser gueldenstaedtii*)

The farms consist of ponds on land, and water is supplied from the rivers that lead to the wide Gironde fjord. This area was previously the natural habitat of the European sturgeon (*Acipenser sturio*), which is now a critically endangered species. In the farms, the fish are fed with a granulate of fish meal, fish oil, plant material, vitamins, and minerals. Research and development of the optimal diet are carried out on an ongoing basis. Groupe KAVIAR was the first caviar producer to use ultrasound scanning to sex sturgeon and determine the maturity of the eggs. This technique is now used worldwide. Sexing can be done after 3 years, and the females must then spend around 8 years in the farms before they are sexually mature and produce eggs.

Before the fish are harvested for caviar production, they are scanned again, and a biopsy is taken to test the taste and maturity of the eggs. They are usually ready in the period from September to March and typically in December–January, when the eggs are at their optimal firmness. At that time, the Siberian sturgeon is around 1 meter long and 7 years old and weighs 7 kilograms, of which 10–12% is eggs. The Russian sturgeon is around 1.5 meters long and 9 years old and weighs 15 kilograms, of which 15% is eggs.

One of the authors (Ole) had in December 2022 the opportunity to speak with Bastien Debeuf, who is responsible for research and development at Groupe KAVIAR. He explains that the company would like to be considered the experts in this niche of caviar production and that the company has a strong focus on technological development that can improve the quality of caviar.

Although the production of caviar at Groupe KAVIAR follows the classical method, there are variations depending on the customer group the product is aimed at. At Groupe KAVIAR, a main quality requirement is a particular firmness of the individual eggs, which on the one hand is determined by the species of sturgeon and the conditions at the farm, but can be influenced by the treatment of the roe for caviar. On the other hand, the color of the eggs is mainly controlled by genetic factors, and the size of the eggs depends on the time of the season and the species of sturgeon.

After the roe sacs have been removed from the slaughtered fish, the roe are sorted by color and egg size. The eggs are then separated from the roe sac and each other by manual mechanical treatment through a special stainless-steel sieve (cf. Fig. 3.5). The separated eggs are subsequently cleaned in fresh water repeatedly, and in this process, damaged eggs and membrane residues can be separated because they will float on top of the water. Due to the osmotic pressure, the eggs grow in fresh water by about 30%. The water used for cleaning is hard and has a high so-called redox level, which helps to suppress microbial activity. This process does not damage the eggs. After being filtered from the cleaning water, the eggs shrink back to their original size.

Bastien Debeuf emphasizes that this entire process is manual and that there may therefore be differences in quality of the product depending on which hands have processed the caviar. It is therefore a matter of know-how and craftsmanship to ensure the quality of the product. At Groupe KAVIAR, this process and the subsequent salting are normally carried out in batches, mixing the eggs from different fish.

Salting is a dry-salting process, which is carried out manually in small buckets on ice, while the ambient temperature is around 8 °C (cf. Fig. 3.6). Not all companies carry out this step on ice. Very pure rock salt is used for salting. During salting, the outer membranes of the eggs are partially modified.

It is the salting process, and not the previous cleaning process, that is mainly responsible for the final firmness of the eggs in the caviar product. The salt percentage can depend on the individual wishes of the customer, but for standard products it is 3.3%, which is typical of the *malossol* technique. Some products also add borax as a preservative in addition to salt.

The caviar is then packed in either smaller vacuum-sealed tins of 10–500 g, which go directly to consumers, or in larger 1 kg or 1.8 kg tins for the wholesaler or for longer-term maturation. This is followed by a period of at least 1 month of maturation of the caviar at a temperature of −4 °C to −2 °C.

Bastien Debeuf says that the caviar market is undergoing major changes under the influence of a large Chinese caviar production. He believes that before the Covid-19 pandemic there was less interest in the country of origin of the caviar, while now there is a lot of focus on origin, especially in top restaurants, and price and quality are no longer the only factors for sales. This is where high-quality French products may gain a market advantage. Bastien Debeuf senses that there has been a rise in caviar consumption, a rise in interest in high-quality products, and a greater focus on the sustainability of caviar production.

Since 2013, French caviar from Aquitaine has been designated with a special label, Caviar d'Aquitaine, which has just been approved as a protected EU regional quality mark (PGI, Protected Geographical Indication) similar to the famous oysters from Marennes d'Oléron.

3.5.4 Color and Egg Size of Caviar

Some might say that there are two colors of the roe you buy in the supermarket. Some are red and yellowish, and some are black, and the black roe are real caviar. However, most people will probably know that the black roe you buy cheaply in the supermarket are not real caviar but from fish whose red and yellowish roe are dyed black. The label and product declaration should also state that it is kaviar or caviar substitute and what raw material the roe are made from. This will typically be from, for example, lumpfish, or perhaps seaweed. The size of the eggs in different types of kaviar substitutes is usually smaller than for real caviar from sturgeon.

However, real caviar is not always black but has a wide spectrum of natural colors, which depends on the species the roe come from, and the color can even vary within the same species and even from fish to fish. The colors can range from jet black or brown to gray to almost golden. The color is an important quality parameter for genuine caviar.

The American shovelnose sturgeon (*Scaphirhynchus platorynchus*) has small eggs that are always jet black. Gray-colored eggs with silver tones are especially typical of beluga (*Huso huso*) but also of sevruga (*Acipenser stellatus*) and American paddlefish (*Polyodon spathula*). The eggs can be both clear and dull. In rare cases, paddlefish produce eggs whose color has greenish nuances. Beluga has the largest eggs, the size of a small pea, and they are soft and creamy.

Black shades toward brown and amber are found in "premium" and "royal" quality caviar from osetra (Russian sturgeon, *Acipenser gueldenstaedtii*; Persian sturgeon, *Acipenser persicus*), kaluga (*Huso dauricus*), and white sturgeon (*Acipenser transmontanus*). The lighter varieties with a golden glow are the rarest and also the most expensive. The eggs are medium sized.

Caviar from sterlet (*Acipenser ruthenus*) consists of smaller eggs, about 1.5 mm, and the color is grayish to brown.

The color of the caviar is to some extent influenced by the treatment with salt, which usually gives a darker color than the "green eggs," which are lighter. In old osetra sturgeons, the eggs retain their light color, and this caviar is particularly sought after and therefore carries a higher price, although the color does not affect the taste. This caviar is called Imperial Caviar.

Finally, in rare cases, you can find white-yellow caviar, so-called Almas (alma = diamond in Persian), from sturgeons that are albinos and therefore cannot pigment their roe. These eggs are often opaque and slightly milky. Only a small proportion of sturgeons are albinos, and their roe are therefore highly sought after and expensive. The same applies to the caviar, which happens to be particularly golden. It is estimated that there are now only very few living specimens in the Caspian Sea of beluga sturgeon that are over 60 years old and that bear the bright Almas caviar.

Aristotle on Caviar
As usual, the Greek philosopher Aristotle (384–322 BC) is said to have been the first to give an authoritative statement on something. Thus, it is also said that Aristotle described caviar as a delicacy that was advertised at banquets accompanied by flowers and trumpets.

3.5.5 Taste, Aroma, and Texture of Caviar

The taste, aroma, and texture of caviar are generally determined by the same factors as described earlier more generally for fish eggs. However, there are some differences that make caviar something special. Firstly, the fat content is high, 10–15%, which gives a particularly creamy mouthfeel. The eggs are also softer than, for example, the somewhat harder trout eggs (which have a similarly high fat content), which further contributes to the creamy taste experience. Finally, it is important to note that caviar products are often far more quality-controlled and specially selected than virtually all other types of fish eggs.

As with other roe products, the salty taste from the processing defines the taste of caviar, just as a certain portion of oxidized fatty acids contributes to the characteristic aroma profile. The umami taste is also largely a taste characteristic of caviar, which is shared with a number of other roe types. The aroma of salted and matured caviar is described as slightly fishy and earthy (from the substance geosmin) with floral notes of cucumber, waxy, and like deep-fried, boiled potatoes. It is substances such as aldehydes, especially heptenal, nonadienal, nonenal, nonanal, decadienals, and methional, that give these aroma impressions. See also the tables at the back of the book (cf. Table 8 at the back of the book).

Traditional uses of caviar involve eating as is, spooning the caviar out of the tin, topping on blinis (cf. Fig. 3.9), and decoration of and condiment to a wide range of dishes including desserts and even chocolate.

Fig. 3.9 Traditional serving of caviar on blinis, here shown along with similar servings of kaviar made by other types of fish roe

3.5.6 *Quality Control of Caviar*

Premium caviar is tightly selected and under strict quality control, both at production facilities (cf. Fig. 3.10) and by authorities. There are a very long series of factors that determine the quality and thus the price of a given type of caviar. Firstly, it depends on the individual fish, its age, size, and health, including its feeding status. The latter, of course, depends on whether it is a wild fish or a farmed fish. Water quality is also important.

Next, the harvesting of the roe plays a role, i.e., the way the roe are removed from the abdomen. Then comes the production method itself, including the amount of salt, pasteurization or not, the temperature conditions under which the roe are stored, the way it is packaged, and finally how long and how the packaged roe have been stored before use. All of these factors determine both the quality and price of

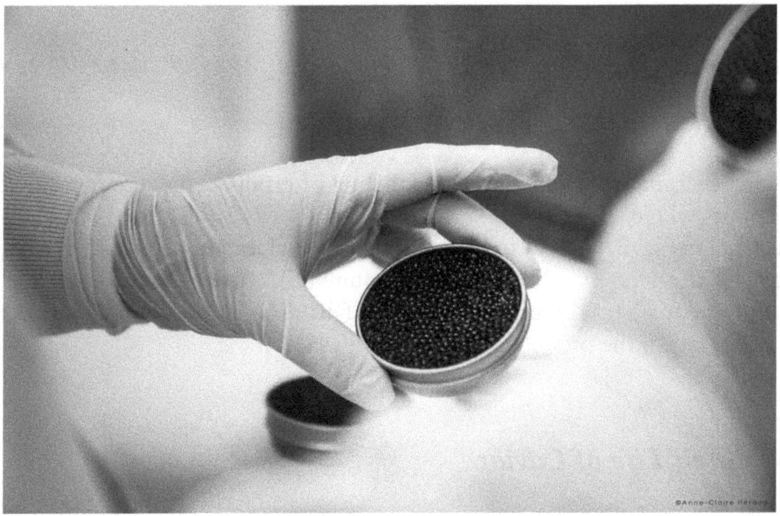

Fig. 3.10 Quality control of caviar

the product. The various caviar producers therefore carry out an extensive program of quality control throughout the process from production to the consumer.

An important step in the quality control and grading of individual caviar products includes assessing the size, color, degree of maturity, uniformity, firmness, clarity, and aroma of the eggs. This assessment is carried out both before and after salting. A given type of caviar is graded into two groups: grade 1 (A), which is the top, and grade 2 (B). Grade 2 typically has smaller and softer, moister eggs. Caviar that falls between the two grades is called select grade. Caviar of average quality or slightly above is marketed as classic and royal, while the highest grades are called supreme or imperial. There are also other, more old-fashioned classifications.

3.5.7 International Registration of Caviar

Since 1998, all species of sturgeon and paddlefish have been listed in an international register under CITES (Convention on International Trade in Endangered Species of Wild Fauna and Flora). To ensure the legality of the international trade in caviar, the world's governments entered into an agreement in 2000 on an internationally standardized labeling system for caviar. In 2004, it was agreed that only caviar that is clearly labeled according to this system can be imported and exported. This applies to all types of caviar, including pressed and pasteurized caviar. The CITES rules presently apply solely to caviar and no other types of fish roe.

These rules mean that all tins of caviar, whether for wholesalers, restaurants, or consumers, have a label with codes describing the species of sturgeon, whether the sturgeon is wild (W) or from aquaculture (C), country of origin, year of production, production company, and batch number. This means that the product is fully

Fig. 3.11 Example of declaration label on tins with caviar (see text for explanation)

```
BAE / C / FR / 2014 / xxxx / yyyy
```

traceable. A similar rule applies to the labeling of products that are repackaged by, for example, wholesalers.

The label in Fig. 3.11 shows as an example the declaration of caviar from Siberian sturgeon (BAE, *Acipenser baerii*) in aquaculture (C), country of origin France (FR), production year (2014), registration code of the production facility (xxxx), and batch code (lot) of the product in question (yyyy).

3.5.8 Shelf Life of Caviar

Caviar should be considered a type of semi-preserved product with a limited shelf life, and temperature control is essential. *Malossol* caviar in an unopened tin in the refrigerator (4–6 °C) can last for about 4–6 weeks without losing flavor or changing texture.

If the tin is frozen and kept airtight, it can last for up to a year. Once the tin is opened, the fats begin to oxidize and become rancid, and the eggs will dry out and within a week have lost their delicate flavor and texture.

Pasteurized caviar can be kept for up to a year in an airtight container, although its flavor and texture will deteriorate. Frozen caviar, whether pasteurized or not, can be kept for a year but will lose quality if thawed and refrozen. Once a tin of caviar is opened, it should be consumed within 3–5 days, preferably sooner. When the tin is opened, it loses its vacuum, and the air that seeps in helps to oxidize the fats in the roe.

Caviar products with more salt than those treated with the *malossol* method can be kept significantly longer, for months or years even without refrigeration.

3.6 Global Caviar Production, Consumption, and Trade

Global production of wild sturgeon was stable at around 20–25,000 tons from around 1950 and peaked at 32,000 tons in 1974. It has since gone down steeply, and from 1984 to 1999, wild sturgeon production fell by almost 90%. At the same time, aquaculture of sturgeon began to gain momentum. Around 2000, aquaculture production of sturgeon reached the same level as wild fish. This level was very low, below 3000 tons worldwide. After that, things moved quickly, and by 2005 aquaculture already accounted for almost 94% of the world market.

In 2021, the legal global market for sturgeon fish was 140,000 tons, of which China accounted for 87%, followed by Russia and Armenia with 4% and 2%, respectively. However, the Armenian sturgeon is only used to a small extent for caviar production. Wild-caught sturgeon accounts for a negligible share. Only a small part of this tonnage of sturgeon is made into caviar. It should be noted that it is very difficult to obtain reliable data for both sturgeon production and caviar production.

A few pure species and hybrids make the largest contribution to the world's caviar production. Siberian sturgeon (*Acipenser baerii*) accounts for the largest part, around 30%, Russian sturgeon (*Acipenser gueldenstaedtii*) for around 20%, and three hybrids of either *Huso* or *Acipenser* together cover around 20%.

Global caviar production in 2021 was 650 tons, of which China accounted for around 250 tons. The largest consumers in the world are the USA, Japan, Russia, and China. The USA is also the largest importer of caviar with around 15% of the world market. In 2021, EU countries produced around 175 tons of caviar, a strong growth of 55% in the period 2015–2018 and 18% since 2019. The majority of European caviar production is due to the four major producers Italy, France, Poland, and Germany, of which Italy accounts for almost a third. France and Spain, in particular, have seen strong growth in caviar production. Europe both exports and imports caviar, with the largest imports coming from China at 65–84% from 2014 to 2018. The modest present caviar production can be compared to the golden days of the mid-1980s, when global production of caviar from wild sturgeon reached almost 3500 tons annually. Annual caviar consumption in Europe is estimated to be around 125 tons in 2018. France has the largest consumption, followed by Germany and Spain.

The rapid growth in caviar production from China has intensified competition in the international caviar market and has led to a decline in market prices, with a 40% drop in prices from 2014 to 2018. Although high-quality Chinese products exist, the market is said to have become muddier due to uncertainty about farming conditions in fish farms and the level of quality control.

Until 2019, there was strong growth in the production of caviar exports from China, and supply began to exceed demand, leading to falling prices and a slowdown in investment in new production sites. The strong growth in the caviar market meant that some types became much more affordable for ordinary people than before. However, the Covid-19 pandemic meant a plunge in caviar production and a reduction in Chinese caviar exports. Due to the uncertain data for global caviar production, it is difficult to make statistically reliable predictions about the future market.

The global production and trade of caviar is regulated by international bodies like FAO, EU, and CITES (Convention on International Trade in Endangered Species of Wild Fauna and Flora). The CITES rules secure absolute traceability of all caviar, although there is a significant black caviar market marred by fraud and organized crime due to the exorbitant prizes on premium quality caviar. Prices are variable, and the most expensive caviar, Almas, is claimed to cost up to $35,000 per kilo. The international market is in a rapid state of change often influenced by fashion and economic crises.

3.6.1 The Price of Caviar

In a way, you could say that what truffles are among mushrooms, caviar is among the different types of kaviar from fish. Caviar occupies a special role in gastronomy, which is due to tradition and not least to the fact that after the decline in wild sturgeon populations, caviar has become a rare and sought-after and thus extremely expensive delicacy. A kilo of the most expensive type of caviar, Iranian beluga, can cost over $30,000. With a little 22-carat gold sprinkled on top, the price is $100,000 per kilo and thus by far the world's most expensive food. But cheaper types of real caviar are also available for about $2000–3000 per kilo. A somewhat cheaper caviar comes from the American paddlefish, which is a very fast-growing species and where a kilo may only cost $1000.

The global market value of legal caviar is difficult to estimate, but analysts believe that it amounted to about $300 million in 2018 and is expected to grow annually by 8–9% and reach $2 billion in 2024. Among the most important factors determining the price of caviar is the species, as some species are rarer and exist in smaller numbers than others. The time it takes for the sturgeon to produce roe (the maturation period) is also a factor, as is the process of removing the roe and preparing the caviar. This includes in particular the thoroughness of quality control and grading of the individual eggs. Finally, supply and demand are a decisive factor, and maturity can also play a role.

The big difference in the price of real caviar compared to kaviar from other types of roe is not only due to the special history and aura that rests on sturgeon roe. It is also due to factors such as the fact that sturgeons grow much more slowly than other fish such as salmon and flying fish, and they do not reach sexual maturity and produce roe until they are much older. In addition, sturgeons only produce roe every other year or even less often. Moreover, surgeons also live in much more limited and specific areas than most other fish, and they are very sensitive to changes in their habitat, such as pollution. On the positive side, the species of sturgeon that are mainly bred in aquaculture are smaller fish, require less food, and can be more quickly driven toward caviar production.

3.6.2 Caviar Counterfeit

The international caviar market is now heavily regulated and monitored by organizations such as the FAO, the EU, and CITES (Convention on International Trade in Endangered Species of Wild Fauna and Flora), and individual countries may have additional specific regulations. As mentioned, the CITES rules ensure full traceability of all legal caviar on the world market.

Both the EU and Europol have identified fishery products, including caviar, as some of the most important products subject to fraud. Interpol is also involved in assessing the illegal catch and trade in caviar, but the possibilities for eliminating illegal activities are limited under the international conventions. There have been examples of smuggling of beluga caviar into the USA, and mislabeled and counterfeit caviar has been discovered. It is now possible to identify counterfeit caviar products more reliably than earlier using molecular biology methods, DNA identification, and PCR techniques that are superior to using traditional studies of egg morphology.

Regulation of the caviar market is nothing new, and the previous classic producers of caviar, Russia and Iran, regulated their markets for centuries. For example, in 1675, the Russian Tsar Alexei forbade the Cossacks from selling their caviar to foreigners. There are examples of the Soviet regime in the 1920s executing producers who did not follow the regime's orders, and the quality of Russian caviar seems to have been stable under these conditions. With the collapse of the Soviet Union, this system also broke down, and the caviar market became to some extent a black and unregulated market, with the result, among other things, that wild beluga became largely extinct and other species very rare. It is unclear whether part of the present alleged Russian caviar production of 40–50 tons per year is an underestimate, because it may include imported caviar from China and Uruguay, which is illegally relabeled in Russia and enters a black market. It is believed that the illegal market for sturgeon and caviar production in Russia could amount to up to $4 billion and thus much larger than the legal one. The Iranians seem to have had a laxer regulation of their fishing waters over time than the Russians.

Despite regulation and international agreements, there is still a significant illegal trade in sturgeon and caviar, which is due to it being a costly product with a high global market value. It is estimated that the illegal trade in caviar has been a decisive factor in the decline of wild sturgeon populations.

On a Field Trip: The World's First Organic Caviar

The company Caviar de Riofrío is located on the cold river Riofrío in the northern part of the Spanish province of Granada. The temperature of the river water is constantly 14–15 °C all year round, and the water is so clean that it can be used directly as drinking water and is bottled by a local producer.

In 2001, the world's first caviar that could be certified as organic was produced here. Almost 40 years had passed before then with trout aquaculture, and sturgeon production began in 1989. Organic certification in particular requires completely clean water, organic feed, and ample space for the fish. It also involves a sustainable and environmentally friendly production method that avoids the use of chemicals, antibiotic, and genetically modified organisms. The facility tries to imitate the fish's natural habitat, and all procedures are carried out to minimize stress, disease, and mortality in the fish.

The main production at Caviar de Riofrío is based on aquaculture of the Adriatic sturgeon (*Acipenser naccarii*), which was originally native to the Adriatic Sea and the Mediterranean Sea, as well as the rivers that flow into the sea from the surrounding countries. This species does not reach sexual maturity until it is 16–18 years old, when it is approximately 1.5 m long and weighs around

20 kg, of which 2–2½ kg can be roe. The Adriatic sturgeon is said to be able to live to be 100 years old and weigh 100 kg. There are only a few wild specimens of the species left, and this sturgeon is therefore a critically endangered species.

Caviar de Riofrío has around 15,000 fish in its holding ponds, of which a few are beluga, osetra, and sevruga. In 2022, 1 ton of caviar was produced from Adriatic sturgeon, whose roe are similar to those of osetra sturgeon. In March 2023, we had the opportunity to visit the company's aquaculture installation in Riofrío and were shown around by Ignacio Alba Alejandre and Laura Cobos Cano, who also conducted a caviar tasting (cf. Fig. 3.12).

Caviar de Riofrío produces three types of caviar, all of which are salted to around 3–3.5%. The first is the organic one, which is only salted, without

Fig. 3.12 Caviar tasting of Adriatic sturgeon (*Acipenser naccarii*) at Caviar de Riofrío in Southern Spain

borax added, and not ripened. The eggs are greenish, quite firm, and very clean in taste. This fine caviar is as close in taste as you can get to fresh roe.

The second product is "Russian-style" caviar, which is ripened for a month and also without borax. The eggs are more gray black with fewer flavor nuances. Finally, there is the so-called traditional version, which involves 4 months of maturation and the addition of borax. The eggs are very dark and creamier, and the taste is somewhat stronger than the other two products. All caviar products are from single fish and not a mixture of eggs from different individuals.

The company has recently launched a completely new caviar product, Alma de caviar, which is dried, salted eggs made from eggs that are either too immature or damaged to be included in standard caviar products. The salt percentage in Alma de caviar is 29%, and the product is suitable for grinding as a flavoring for other dishes. The taste is very fine and pure and unmistakably caviar.

We ended our visit to Riofrío by having lunch in one of the local village's three restaurants, located along the river in the town square. They serve various dishes with sturgeon meat, fried, dried, and smoked. The meat for this comes either from the males, which are killed on the sturgeon farm at the earliest at the age of 8, after being sexed, or from the females slaughtered after they have delivered caviar. Nothing goes to waste.

Recipe 3.1 Raw-Marinated Norway Lobster with Roe and *Ponzu* (Fig. 3.13)
Per person as a small appetizer

1 large or 2 smaller fresh Norway lobsters
7% brine, i.e., ½ l water and 35 g salt
1 tbsp kaviar, preferably several different types
½ tbsp good quality *ponzu*
A little lime juice

1. Make a 7% brine and cool it down to 5 °C.
2. Peel the meat from the Norway lobsters and carefully remove the guts.
3. Place the Norway lobsters in the cold brine for 7 min, remove them, and let them drain well on a clean cloth or absorbent paper.
4. Cut the Norway lobsters into slices and place them in a deep plate.
5. Mix some lime with *ponzu*, drizzle the juice on top, and spread the roe.

Fig. 3.13 Raw marinated Norway lobster with caviar (here herring roe and caviar) and *ponzu*

Chapter 4
Composition of Fish Roe

Fish roe can be considered a nutritious food, both in terms of macronutrients such as protein and fat as well as minerals and vitamins. There is of course variation for different species, just as the amount of the different nutrients depends on the season and the degree of maturation of the eggs. The content also depends on the fish's feeding status, whether they are from saltwater or fresh water, and whether they are wild or farmed on a specific type of feed.

The variations in the composition of the roe apply in particular to the unsaturated fatty acids and to a lesser extent the amino acids. The fatty acids from the food are incorporated particularly effectively in the fish's ovaries and thus in the eggs to a much greater extent than in the muscle tissue, which reflects that the unsaturated fatty acids are particularly important for reproduction. The difference between the content of unsaturated fatty acids in the roe of freshwater fish and saltwater fish also reflects that access to the polyunsaturated fatty acids is much richer in salty marine environments. Many farmed fish are fed with feed containing vegetable oils, which is manifested in a shift toward a higher content of ω-6 fats in their roe.

The water content of fresh fish roe is typically around 70% and the protein content around 25%. The fat content varies from 5% to 20%. Please refer to the tables at the end of the book for details. Fish roe are rich in vitamins such as A, B_1 (thiamine), B_2 (riboflavin), B_5, B_6 (folate), B_{12}, D_3 (calciferol), E, and K, and there are particularly high amounts of vitamin B_2, vitamin D, and vitamin B_{12}. The mineral content (ash) is typically 1–2%, and this concerns mainly calcium, iron, magnesium, phosphorus, sodium, and selenium.

Fish roe typically contains very little carbohydrate (cf. see Table 1 at the end of the book), which reflects the fact that the development of fish larvae is based on energy from fat and protein alone. When it comes to the nutritional value of proteins and fats, it is the distribution of different amino acids and fatty acids, respectively,

O. G. Mouritsen, K. Styrbæk, *Roe and Roe Gastronomy*,
https://doi.org/10.1007/978-3-032-13142-3_4

that is important. For amino acids, it is crucial that they include the so-called essential amino acids, which we must obtain from the diet and which our body cannot produce itself. Regarding fats, fish roe are particularly rich in the so-called super-unsaturated ω-3 and ω-6 fatty acids, which are also essential and which we cannot produce ourselves.

The composition and amounts of both amino acids and fatty acids in roe are different from those in muscle meat, which demonstrates that the two types of tissue serve very different purposes. Muscles must provide strength and movement, while roe must provide nutrients and energy for the growth of fish larvae before and after hatching.

The content of protein and the profile of fatty acids depend on the fish's food and thus on whether the fish is wild or farmed. Consequently, a higher content of amino acids in the feed results in a higher content of protein in the roe, and in particular the content of the super-unsaturated fatty acids in the feed is reflected in the fat content of the roe. This makes it possible to influence the important ω-3/ω-6 ratio, which can often be low in roe from farmed fish if their feed contains a lot of vegetable oil, which especially contains ω-6 fats.

For details regarding the different composition and energy content of roe from different fish species, please refer to the tables at the back of the book.

4.1 Proteins and Amino Acids

The proteins in fish roe contain all the essential amino acids (cf. see Table 3 at end of the book). The protein content depends to a large extent on the degree of maturation of the eggs. Although the amino acid composition varies for the different types of roe, it is usually dominated by glutamic acid, which is a nonessential amino acid. This is followed by aspartic acid, leucine, and lysine. Glutamic acid in the form of free glutamate is absolutely crucial for the umami taste of the roe.

As mentioned, fish species are often divided into so-called demersal and pelagic fish, corresponding to species that predominantly reside on the bottom of the sea and higher up in the open water masses, respectively. It turns out that the content of free amino acids is relatively higher in eggs from pelagic fish. While around up to 50% of the amino acids are bound in proteins in the roe of pelagic fish, the corresponding figure is 92–95% for demersal fish and freshwater fish. One might therefore expect that roe from pelagic fish has the most umami taste. However, this does not always have to be the case, as there is a large variation for the different amino acids.

4.2 Fats

With regard to the fat content (cf. see Tables 1 and 6 at the end of the book), the different types of fish roe can be roughly divided into two groups. Some (e.g., sturgeon, salmon, trout, and pike) have more than 12% fat, and others (e.g., herring, cod, and flatfish) have less than 5% fat. The fat content reflects to some extent the natural ecological habitat in which the fish live. Fish that spawn in nutrient-poor water have eggs with a high nutritional content of fats and often also protein. In contrast, fish that spawn in nutrient-rich water such as salty marine seas typically have smaller amounts of nutrients packed into the eggs but enough to hatch the egg. As mentioned earlier, it is mainly the unsaturated fats that characterize the fats in fish roe. About 30% of the fat is saturated, so there is a predominance of unsaturated fatty acids and especially large amounts of the super-unsaturated ω-3 (DHA and EPA) and to a lesser extent ω-6 fats (AA).

The ω-3/ω-6 ratio between the content of ω-3 and ω-6 fats in roe varies greatly from species to species (between 1 and 30) but is typically around 5, suggesting that fish roe are an exceptionally rich and healthy source of unsaturated fats.

There is a clear difference between the fatty acids in roe from saltwater fish and freshwater fish, and the content of ω-3 fats is higher in eggs from saltwater fish. This reflects the fact that the availability of ω-3 fatty acids is more limited in freshwater environments. Fats are found in many different chemical forms, some as phospholipids and others as triglycerides. For example, cod roe contains 60% of the fat in the form of phospholipids, and 8–12% is triglycerides. Salmon roe contains equal amounts of both types of fat.

The high content of polyunsaturated fatty acids, especially the super-unsaturated fatty acids EPA, DHA, and AA, is the most important nutritional advantage of fish eggs (cf. see Table 6 at the end of the book). There is considerable variation between species and some dependence on the fish's diet. The content of vegetable oils in the feed for fish in aquaculture is reflected in a higher content of ω-6 fatty acids.

There is a large difference in the fat composition between the fish's muscles and eggs, especially during the maturation of the eggs, where there is an accumulation of the super-unsaturated fatty acids (DHA, EPA, and AA), and the saturated and monounsaturated fats as well as the short-chain ω-6 fatty acids are converted by metabolic activity in connection with the formation of the eggs. The importance of the super-unsaturated fats for the reproductive purposes of the eggs and the development of the embryo is emphasized by the fact that the content of these fatty acids in the egg is only slightly dependent on the fish's diet compared to that in muscle tissue.

The fat composition of the individual types of roe does not seem to be affected by the usual preparation methods, and, for example, "green" sturgeon eggs and caviar have the same fat composition.

A very significant difference between eggs from fish and eggs from land animals such as chicken eggs is that the fat content in chicken eggs is typically twice as high, and the composition of fatty acids in the fat is completely different. For details, see Table 6 at the end of the book. This difference is not only reflected in the important ratio ω-3/ω-6, which is approximately 5 for fish eggs and only 0.13 for chicken eggs (cf. see Table 1 at the end of the book), but also in the fact that some of the fatty acids in fish eggs have an odd number of double bonds, whereas the fatty acids in chicken eggs only have an even number of double bonds.

A very special fat is cholesterol. Cholesterol is an important substance for building cells, and it also functions in the formation of hormones and bile salts. The cholesterol content is particularly high in roe, where it is typically ten times higher than in the fish's muscles. The cholesterol content is generally around 0.3% but varies somewhat from species to species and is somewhat lower than the content in chicken eggs (approximately 1% in egg yolk) (cf. see Table 1 at the end of the book). The cholesterol content is particularly high in salmon roe and sturgeon roe. Again, reference is made to the tables at the back of the book (cf. see Table 1 at the end of the book).

The Important Super-unsaturated Fatty Acids: ω-3 and ω-6

Fatty acids are divided into saturated and unsaturated fatty acids. The unsaturated ones can have different degrees of unsaturation, which is described by the number of double bonds in the hydrocarbon chain of the fatty acid. Unsaturated fatty acids have one or more double bonds. Fish roe are characterized by having a high content of polyunsaturated fatty acids, especially the kind that has many double bonds on a fairly long hydrocarbon chain. Here, the unsaturated fatty acids, which have three or up to six double bonds, play a very special role. There are two classes in particular, which are referred to as ω-3 and ω-6 fatty acids, respectively. Important fatty acids in the ω-3 class are the super-unsaturated eicosapentaenoic acid (EPA, with 20 carbon atoms and five double bonds) and docosahexaenoic acid (DHA, with 22 carbon atoms and six double bonds). In the ω-6 class, the most important super-unsaturated fatty acid is arachidonic acid (AA, with 20 carbon atoms and four double bonds). The ratio, ω-3/ω-6, between the amount of ω-3 and the amount of ω-6 fatty acids plays a significant role for the health value of foods containing fats, and a high value of this ratio is preferable. Fish roe generally have very high values of the ω-3/ω-6 ratio, typically around 5 or higher.

4.3 Energy Content

The energy content of fish roe is somewhat higher than in the fish's muscles, but it depends a lot on the type of roe, because the fat content can vary from 4% in cod roe to up to 20% in salmon roe and caviar. The energy in fish roe comes from fat and protein, as there are no or only very small amounts of carbohydrates in fish eggs.

Compared to chicken eggs, only the roe of fatty fish comes close in calorie content. For example, salmon and sturgeon roe have about the same energy content as chicken eggs (1340 kJ/100 g), whereas cod, lumpfish, and herring roe only have about a third of this (cf. see Table 7 at the end of the book).

The content of a chicken egg corresponds to about three tablespoons of roe (45 ml) of, for example, caviar. In such a serving, caviar contains 120 kcal, 9 g fat, 0.28 g cholesterol, and 0.7 g sodium. In contrast, a chicken egg contains only 75 kcal, 5 g fat, 0.21 g cholesterol, and 0.6 g sodium.

Recipe 4.1 Roe Mayonnaise on Chicken Eggs (Fig. 4.1)
Serves 4

Approx. 100 g frozen and thawed roe* or kaviar in a jar
4 chicken eggs
1½ g salt
1 tsp mustard
2 tsp white wine vinegar
2 tsp lemon juice
1 pinch cayenne pepper
Approx. 2 dl neutral oil
A little cress

1. Blend the roe, salt, white wine vinegar, lemon, and mustard well together with a hand blender.
2. Add the oil in a quiet, continuous stream and blend; stop when the mayonnaise is thick and creamy.
3. Flavor with lemon, salt, and cayenne pepper.
4. Cook the chicken eggs until hard-boiled and refrigerate until needed.

The roe mayonnaise is good for a small starter, e.g., for boiled eggs with cress, or as a base for dips, dressings, and creams for various fish dishes. The roe acts as an emulsifier here, and it is therefore important that the fish eggs are blended into pieces.

** The roe are used raw and, therefore, it must be frozen at least 24 hours before use.*

Fig. 4.1 Chicken eggs with cream (roe mayonnaise) made from fish eggs (here herring roe)

Chapter 5
Fish Roe from All over the World

There are at least 40 different types of fish roe used as kaviar, and the global market amounts to up to 60,000 tons, almost 100 times the production of caviar from sturgeon. However, some of these types of roe are valued in their own right and are not thought of by consumers as a caviar substitute. The market for these types of roe is growing, partly due to the limitations in the production of real caviar, and sales have grown sharply, especially since the 1960s. Another factor is the globalization of the sushi culture, where roe from fish such as salmon, trout, capelin, and flying fish have come to the attention of consumers as a product with an interesting taste and texture that comes into its own in special applications. A list of some typical gastronomic uses of roe is found in Table 11 at the end of this book.

Some types of roe traditionally have their own unique and time-honored names around the world. Caviar from sturgeon naturally occupies a special status. In the Nordic countries, this is, for example, *lögrom* from vendace or *sikrom* from whitefish. In Japan, flying fish roe (*tobiko*) has a very special status, and the whole gonads, *uni*, from sea urchins are considered a special delicacy. Roe from various fish, especially salmon, are called *ikra* in Russian, and the Russians were the first to start processing salmon roe into loose eggs as a substitute for caviar in the 1830s. The Russian term is found in the Japanese name, *ikura*, which is also recognized as an important gastronomic product in Japan. Finally, a roe product from antiquity, *bottarga*, can be mentioned, in the form of dried whole roe sacs from, for example, mullet.

Unlike caviar, most types of fish roe and kaviar are valued more for their texture than their taste. In particular, the ability of some types of roe to be crunchy is their special advantage. Many roe products from several of the fish species described below are designated by a Japanese name, which clearly expresses the prominence of the Japanese in the use of fish roe for various gastronomic products.

O. G. Mouritsen, K. Styrbæk, *Roe and Roe Gastronomy*,
https://doi.org/10.1007/978-3-032-13142-3_5

***Tarama*: Fish Roe in Greek**

Tarama means fish eggs in Greek, and the word is used in connection with a special mixture of fish roe, especially from cod and other codfish, but also carp, herring, halibut, and sometimes salmon and sea urchin roe in various parts of the world. The mixture, called *taramasalata*, consists of roe (which may be smoked) and oil, to which bread or mashed potatoes, lemon juice, and spices such as pepper and garlic have been added. The roe content in a *taramasalata* varies from 20% to 50%. *Taramasalata* is especially popular in Greece and France.

Recipe 5.1 *Taramasalata* (Fig. 5.1)

100 g frozen and thawed cod, mullet, or flatfish roe*

¼ tsp salt
50 g shallots
1 clove of garlic
Zest and 30–40 g juice of 1–2 unpeeled lemons
100 g sandwich bread without crust
2½-3 dl neutral oil, e.g. rapeseed or sunflower seed oil
120 g ripe, blended and drained fresh tomato juice
1 tsp sugar
1 pinch cayenne pepper
2 tbsp olive oil
A little more olive oil for serving

1. Chop the tomato in a mini chopper and sieve out the 120 g that will be used. Save the pulp.
2. Pour the juice back into the mini chopper and blend the onion in it.
3. Grate lemon zest into it, add roe and ¼ dl lemon juice, press garlic into it, grate and place sandwich bread without crust in it, and add sugar and a little cayenne pepper.
4. Turn on the mini chopper and add the oil in a thin stream.
5. Blend/stir until smooth, and season with salt, pepper and lemon juice.
6. Serve in a wide small bowl and pour a little olive oil over it.

**If using raw roe, it must be frozen for at least 24 hours before use. Canned roe or kaviar from a jar can also be used.*

Taramasalata is a dip that can be used like hummus with, for example, roe crackers or as a starter with bread. If you want a more rustic version of *taramasalata*, you can add tomato pulp, a little finely chopped onion and roe, which are not blended, but just stirred in at the end.

Fig. 5.1 *Taramasalata*

Below we review alphabetically different fish species, some of which yield very well-known and widely used roe products, e.g., from cod, salmon, and lumpfish, while others are more unknown or only rarely used. Caviar from sturgeon was discussed in its own chapter, Chap. 3. We have also included a few fish species whose roe have only a small tradition of being used and where in many cases they are not a commercial commodity at all but could well be if there was a demand, e.g., roe from flatfish. The tables at the back of the book contain data on the ingredients in the roe of the fish mentioned below, in addition to a number of others (cf. Table 1 at the end of this book).

5.1 Burbot

Burbot (*Lota lota*) is a codfish, whose roe are somewhat confusingly also referred to as *sikrom*, like the roe of the whitefish. It is the only type of codfish that lives in fresh water. Up to 25% of the fish's weight can be made up of roe just before spawning. The roe are used for various kaviar products, just like roe from herring. The roe have a high content of waxy fats. The eggs are 1.2–1.4 mm in diameter.

5.2 Capelin

Capelin (*Mallotus villosus*) is a fish in the smelt family (Osmeridae). Its eggs have a light orange-yellow color and are very small (1.0–1.2 mm) and slightly smaller than eggs from flying fish roe (*tobiko*). The eggs have a crunchy mouthfeel that is comparable but not as crunchy as *tobiko*. The eggs are also marketed as Arctic Sea roe, reflecting that capelin lives in the cold Arctic seas around Greenland and Canada. Capelin roe are particularly rich in ω-3 fats, and the ω-3/ω-6 ratio is high (15–20).

Capelin roe are called *masago* in Japanese and are often used as an imitation or substitute for the rarer and more expensive *tobiko*, where the eggs are slightly less bitter. In lightly salted form, *masago*, like *tobiko*, is primarily used as a garnish and decoration for food, e.g., in sushi preparations. In the same way as *tobiko*, capelin roe are sometimes colored and flavored with *yuzu*, *wasabi*, chili, and squid ink.

5.3 Carp

Carp (*Cyprinus* spp.) is a freshwater fish that is widespread as food fish, e.g., in Central Europe, where it is raised in fish farms. The roe are often used in, e.g., *taramasalata*. The eggs are 0.8–1.6 mm.

A very special preparation of carp roe is in the form of the traditional *funa-zushi* made from Japanese crucian carp or nigoro-buna (*Carassius auratus grandoculis*) whose habitat is limited to Lake Biwa near Kyoto in Japan. In *funa-zushi*, the freshly caught carp, brimming with roe, is fermented in rice for months (cf. Fig. 5.2).

5.4 Cod

Cod covers various species of the genus *Gadus*, e.g., Atlantic cod (*Gadus morhua*) and Pacific cod (*Gadus macrocephalus*). The eggs are typically 1.3–1.4 mm and contain 20–25% protein and 2–5% fat. The ω-3/ω-6 ratio is large and in the range of 10–15. The annual production of cod roe is around 5000 tons, and Iceland, Norway, and Denmark are the world's leading producers.

Fig. 5.2 Japanese crucian carp prepared as *funa-zushi* where the fish is filled with roe. Kindly prepared by Professor Ishida Masayoshi

Cod roe are a traditional and very popular food in the form of whole roe sacs, which are either salted, marinated, smoked, grilled, or fried, and the roe are eaten both hot and cold. Loose, salted eggs find their way into dressings and spreads, where the roe are, for example, mixed with cream cheese in a product that can be spread on bread and sold in tubes ready to use. Roe products from other codfish, such as pollock (Alaska pollock), are often somewhat misleadingly declared as originating from cod.

5.4.1 Salted Cod Roe

Roe from codfish, and especially Alaska pollock (*Gadus chalcogrammus*), is the basis for two Japanese specialties, *tarako* and *mentaiko*, that are based on salted cod roe (cf. Fig. 5.3). *Tarako* simply means cod roe in Japanese (*tara* is cod and *ko*

Fig. 5.3 Top: Salted Alaska pollock roe (*tarako*). Bottom: Spicy salted Alaska pollock roe (*menteiko*)

means children), and roe from both pollock and cod are used for classic *tarako*. *Mentaiko* (or *karashi-mentaiko*) is a spicy variant of *tarako*. A wide variety of more or less strong seasonings can be used, for example, garlic or Japanese chili (*togarashi*).

The salting is simply done by placing the fresh roe sacs in saltwater (12–20% by weight) and draining the water after a few days. To achieve the best consistency, at least 16% salt is needed, but it can also depend on whether other salts or additives are used in the production.

Salted cod roe are popular in South Korea and Japan. In fact, *tarako* is originally a Korean invention, believed to date back to the late fourteenth century and only became popular in Japan in the late 1940s. In Japan, salted cod roe in the form of *tarako* is widely consumed, usually based on Alaska pollock, although the product is often declared as coming from cod. The salted *tarako* can be marinated with various spices and then becomes *mentaiko*. *Mentaiko* is used both with and without the membrane of the roe sac, which has a slightly tough texture, but can be easily removed with the fingers, or the salted roe can be pressed out of the sac. *Tarako* has a neutral, slightly pink color, the mouthfeel is tender, and the taste is fairly neutral, before it is possibly marinated into *mentaiko*. The green spots that can sometimes appear on the roe are due to remnants from the fish's bile. The color of *mentaiko* depends on the spices used. *Tarako* and *mentaiko* can be kept for 2–3 days in the refrigerator and 2–3 months in the freezer. *Tarako* is naturally salty, and unseasoned raw *tarako* has a mild sea flavor and no fishy aroma. When *tarako* is heated, the identity of the individual eggs dissolves, and the roe mass becomes a creamier paste.

In Japanese cuisine, *tarako* is eaten as it is or grilled (*yaki-tarako*). Like *tataki*, the roe are cooked on the outside but remain raw in the middle. *Tarako* is also used as a filling in rice balls (*onigiri*) and on top of cooked rice, among other things, to provide umami taste. In addition, *tarako* is consumed with pasta and noodle dishes. Another traditional use is *tarako* folded in a *shiso* leaf, breaded, and prepared as *tempura*.

Mentaiko often accompanies boiled rice or is mixed into sauces for pasta dishes and enters salads and omelettes. Consumption of *mentaiko* is about twice that of plain *tarako*. *Karashi-mentaiko* is often used as a popular souvenir. An inexpensive spicy variant of *mentaiko* called spicy roe (*metai*) is produced in the USA for export to the South Korean market, where it is used as a condiment with added nitrate. Salted cod roe are used, like roe from other fish, in *taramasalata*. Salted cod roe also find uses in Russian and French cuisine. Unlike other salted and preserved roe products, such as mullet roe in the form of *bottarga*, the umami taste of salted cod roe decreases during processing. Thus, the equivalent umami concentration (EUC) of Alaska pollock roe decreases from 8400 mg/100 g to 1800 mg/100 g when the roe are processed into *tarako*. Yet it is the content of free amino acids and nucleotides that, together with the high salt content, determine the taste profile of *tarako*.

5.4.2 Canned Cod Roe: Not Only from Cod

Canned cod roe are a widespread product and used, for example, on open sandwiches or for frying (cf. Fig. 5.4). Cod roe are popular in the same way as mackerel and are to that extent canned fish for the public, without necessarily compromising on gastronomic quality. The popularity is undoubtedly due to the roe having an inviting, slightly crunchy texture and lots of umami taste, which is also enhanced by the fact that the producer has added tomato concentrate. Originally, the tomato concentrate was added to adjust the slightly dull gray color of the roe, but the fact that glutamate from tomatoes provides umami synergy together with the nucleotides in the roe must also have played a role. The product also has a fairly neutral aroma, so it does not "smell of fish" in any way.

Fig. 5.4 Canned cod roe and fried cod roe with remoulade on rye bread

Often there is not only cod roe in the cans but a mixture of cod roe with roe from, for example, flatfish such as plaice and dab as well as from other codfish such as pollock and blue grenadier (*hoki*; *Macruronus novaezelandiae*). The color of the product is light brown with a pinkish tinge, which is due, among other things, to the addition of tomato concentrate. The roe are also added with salt and rapeseed oil. The roe mass is heat-treated in the can together with thickeners such as carrageenan, locust bean gum, guar gum, or xanthan gum and thus becomes so firm that it can be cut into neat slices without breaking.

There is a big difference between different canned cod roe products. Some contain around 75% roe, while there are others with only 55–65% roe, which, as mentioned, is not always from cod. Cod roe mixed with other types of roe are not of inferior quality to that made from pure cod roe. In order to obtain a uniform firmness of the canned product, the manufacturer regulates the total dry matter content. Roe from pure cod would be too dry and firm without this modification, and strangely enough it is the roe that, based on the dry matter content, would be assigned the lowest quality that is actually best suited for making canned cod roe. Canned products with over 90% cod roe would be inedible, as they are too firm and dry.

There are organic versions of canned cod roe. Here, the roe are from wild-caught cod with MSC certification and are therefore sustainable, although not truly organic. However, the product is supplemented with organic rapeseed oil and organic tomato concentrate.

At Amanda Seafoods A/S in Frederikshavn, the now legendary Amanda cod roe are produced. The product is based on frozen cod roe, which is brought in from the sea, possibly together with other roe. Cod roe are used from both Atlantic cod (Norwegian *skrei*) and Pacific cod. The cod fish can weigh anything from 2 kg to 40 kg. The frozen roe are stored for six months to 1 year and thawed from −24 °C to −2 °C in 20 minutes in the factory in a radio-wave oven, which ensures uniform and rapid thawing. By passing the thawed roe through special rotating sieves, the loose eggs are pressed out of the roe sacs, and the membranes are separated out. These membranes are now completely crushed and returned to the fluid mass of eggs. This is an important step that ensures that the roe mass is still liquid and that the slightly bitter taste of the membranes is preserved in the product. The bitter aftertaste of cod roe is one of the quality parameters. The mass is now added with tomato paste and the necessary amount of thickener, such as carrageenan, to obtain the correct consistency. The recipe varies only a fraction from the original one, which Amanda's founder, Finn Tengberg-Hansen, developed in Kerteminde, a small coastal town in Denmark, in the 1950s. One thousand liters of roe mass is processed at a time, and the factory can produce 400 cans per minute of various sizes and shapes. After the cans are sealed, they are passed through an autoclave, which is a closed tank, where the cans are heated to 108–116 °C with the help of steam under a pressure of two bars for a couple of hours. As a special trick, the liquid cod roe are filled into the cans in such a way that during the autoclaving process a thin layer of carrageenan is formed on the inside surface of the cans. This layer acts as a sliding surface when the consumer has to get the roe out of the can, and the layer can be easily rinsed off with water. The standard shelf life of the canned product is set at 5 years, but in practice it is somewhat longer.

5.4.3 Cod Roe as a Spread

A popular kaviar product on the Scandinavian market is roe in a tube (cf. Fig. 5.5), and it is the Swedes in particular who are major consumers of this product. The content, which can be easily spread on bread, typically consists of salted cod roe with a little sugar. The roe may also be smoked. A high salt content means that the product does not need to be pasteurized and can be kept refrigerated for a long time. Roe in a tube is popular with children, and a Norwegian manufacturer expresses this in an advertisement as follows: "Mom says it's healthy, but funnily enough, I still like it."

The Swedish product Kalles Kaviar has an iconic status in Swedish folk cuisine. It is cod roe in a tube, and it is used as a spread on bread. The product contains 77% roe, sugar, potato slices, onion, tomato paste, rapeseed oil, lactic acid, spices, and

Fig. 5.5 Kalles Kaviar: roe as spread in a tube

preservatives. The tomato paste content indicates that the manufacturer is probably aware of the taste effect of the umami synergy between the roe's nucleotide content and the tomato paste's content of glutamate.

On a Field Trip: Amanda Cod Roe—A Legend

Some people know the song about the fishing girl Amanda and the Funen market town of Kerteminde. That was probably why director Finn Tengberg-Hansen (1913–1991), son of the famous industrialist Chr. Hansen, at the Konservesfabrikken 555 in Kerteminde in 1957 chose the name Amanda for a new canned fish product made from cod roe. Until then, you could only get whole small cod roe sacs in cans. Here, the roe was placed in saltwater, and the saltwater tended to splash out when you opened the cans. Only a little of all the good cod roe from large cod could therefore be canned industrially.

But in the 1950s, 555 had figured out how to remove the membranes from the roe sacs from large cod and make the mass of cod eggs so fluid that it could be pumped through pipes. The cans could then be filled completely with roe before being preserved by autoclaving (boiling). To give the roe a suitable firmness, a thickener of a seaweed extract (carrageenan) was added. So that the consumer could easily get the roe out of the can whole, the inside was coated with a varnish so that the solid mass of cod roe could easily escape. The roe was also shaped into a block so that it could be cut into pieces that would fit half a slice of bread.

Amanda cod roe became a great commercial success both in Denmark and internationally. In 1974, production moved from Kerteminde to Munkebo under the name Munkebo Fine Foods, later Munkebo Seafoods A/S. In 2017, 555 was sold to a Norwegian company, 101 years after the founding of the company, which before the cod roe adventure had made canned fish balls and canned boneless herring. Amanda cod roe are today produced at Amanda Seafoods A/S, which is Denmark's largest producer of cod roe and is based in Frederikshavn. Amanda is now one of only three Danish fish canning factories, where there were around 100 in the 1970s.

We visited Amanda in January 2023, where product development manager Ronnie Nielsen and marketing manager Brian Gert showed us around. We saw how the frozen roe that come in from the sea are turned into loose roe and how the product is turned into the well-known solid blocks in cans (see Fig. 5.6). All Danish cod roe are now produced at Amanda, although some of it appears under "private labels" in the retail trade. Products are made from pure cod roe and some mixed with roe from other fish. This amounts to around nine million cans per year. The market is reasonably stable in Denmark, and Amanda is working on developing special convenience products for the English market.

Fig. 5.6 Canning of cod roe

Recipe 5.2 Klavs' Kaviar (Fig. 5.7)
Serves 4–6

200 g roe from flatfish or cod*
½ dl white wine
1½ tsp onion powder
2 tbsp ketchup
2 tsp lemon juice
30 g capers
1 tsp Worcestershire sauce
1 dl neutral oil, e.g. grapeseed or rapeseed oil
Freshly ground pepper and salt

Fig. 5.7 Klavs' kaviar

1. Make an incision lengthwise in the roe sac and carefully scrape out the eggs.
2. Pour white wine into a pan, heat, add the roe, and season with a little salt.
3. Put the lid on and turn the roe a few times with a dish scraper; it only needs to be cooked for approx. 2–3 min, until it is cooked through. Cool the roe in the refrigerator.
4. Put the roe in a mini chopper with onion powder, ketchup, lemon juice, capers and Worcestershire sauce, blend, and add the oil a little at a time in a thin stream while turning. Add a little extra oil if you want your Klavs' Kaviar a little thicker.
5. Taste and refrigerate.
6. Spread the cream on rye bread, crispbread or chips, or use it as a dip for crispy crackers such as roe crackers.

The recipe is an example of how you can very easily make your own "Kalles Kaviar" and adjust the taste as you please.

**Canned cod roe can also be used.*

5.5 Flatfish

Roe from flatfish are discussed here collectively, as they have a number of common features. Although flatfish are important food fish, their roe are not a common commodity. However, there is nothing wrong with the roe, either in terms of taste or possible gastronomic uses, but they are either cut off at the fishmonger or not eaten when the consumer prepares a whole fish that happens to have roe. This applies, for example, to roe from plaice, flounder, dab, brill, and turbot. From the rest of the Atlantic area, it can be halibut and halibut. Flatfish are divided into right-handed flatfish belonging to the flounder family, also called plaice family (Pleuronectidae), and the left-handed belonging to the turbot family (Scophthalmidae).

In addition to plaice (*Pleuronectes platessa*), the plaice family also includes Greenland halibut (*Reinhardtius hippoglossoides*), Atlantic halibut (*Hippoglossus hippoglossus*), dab (*Limanda limanda*), flounder (*Platichthys flesus*), witch (*Glyptocephalus cynoglossus*), sole (*Solea solea*), and lemon sole (*Microstomus kitt*). The turbot family includes turbot (*Scophthalmus maximus*) and brill (*Scophthalmus rhombus*).

Generally, there is little information about the composition of roe from flatfish, but they are predominantly roe types with a large water content and only a little fat. The protein content varies quite a bit. The eggs are small, around 1–1.5 mm. For the species where the fats have been studied, it turns out that the ω-3/ω-6 ratio is higher than 5.

Before spawning, flatfish such as flounder can contain very large roe sacs that have displaced a significant part of the fish muscle, which is then loose and not so suitable for eating. The roe, on the other hand, is a delicacy and should be more appreciated. A lot of flatfish roe ends up as a filling in industrially produced cod roe products. When the eggs are close to maturity, they can be easily scraped out of the roe sac with a spoon and, after a light and brief salting, can make excellent kaviar. Alternatively, you can boil, steam, or fry the roe sacs or cook them together with a whole fish. Flatfish roe from, e.g., turbot is excellent as an emulsifier in mayonnaise instead of egg yolk.

5.6 Flying Fish

Flying fish constitute a large family (Exocoetidae) of marine fish, e.g., *Cheilopogon furcatus*, *Cypselurus opisthopus hiraii*, and *Exocoetus* spp., which produce very small eggs, 0.5–0.8 mm, which are golden to orange in color. The eggs contain 10–12% protein and 2–3% fat. The eggs are extremely crispy and crunchy, and they are prepared like other types of roe in the form of loose eggs with salt. In Japan, they are known as *tobiko*, and it is popular to both color and flavor *tobiko* with squid ink, *yuzu*, *wasabi*, soy sauce, or chili. *Tobiko* is used in sushi, e.g., *gunkan*-sushi (battle-ship sushi), or as a topping (cf. Fig. 5.8) or filling in *maki-sushi* (rolled sushi).

Due to price and limited availability, *tobiko* is often imitated with herring roe or capelin roe (*masago*). Whereas herring roe can have the same crunchy mouthfeel as *tobiko*, *masago* is softer and less crunchy. *Tobiko* has a salty and slightly smoky flavor and is sweeter than salted salmon and sturgeon roe. The unique crunchy mouthfeel of flying fish eggs makes them suitable for adding an interesting extra sensory component to various green dishes.

Fig. 5.8 *Gunkan-sushi* with roe from flying fish (*tobiko*)

5.7 Garfish

Garfish (*Belone belone*) has roe with relatively large eggs (3.0–3.5 mm), which are pale red. Since garfish is often sold as whole, freshly caught fish, it is not unusual to get it with roe. The roe can easily be used lightly salted as kaviar.

5.8 Haddock

Haddock (*Melanogrammus aeglefinus*) is native to the cod family. Its roe are rarely found on the market but can be used for the same purpose as cod roe. The eggs, like those of other codfish, are around 1.2–1.4 mm in diameter.

5.9 Hake

Hake (*Merluccius merluccius*) has roe that are similar to that of the codfish, to which it is related. The eggs are relatively large, about 3 mm in diameter. Hake roe are popular in Spain, especially in Andalusia, where the roe are typically fried. Fresh hake roe are eaten in Portugal with a little olive oil and vinegar, and it is also found on bread in tapas bars in Spain. In Uruguay, boiled hake roe are a traditional dish in the fall. The roe have a high content of waxy fats.

5.10 Herring

Herring (*Clupea harengus*) is a schooling fish that lives in the North Atlantic and is therefore also called Atlantic herring. Another species of herring (*Clupea pallasii*) is found in the Pacific Ocean. There are a large number of local herring populations, and in the Nordic countries the herring reaches both the Baltic Sea and the Gulf of Bothnia. The herring's roe sacs are approximately 3 cm wide and 8 cm long and consist of rather small eggs, 1.3–1.5 mm, which are whitish yellow. When spawning, the herring roe can constitute 10–20% of the body weight.

The different strains of herring have different spawning times, some in spring and others in autumn and winter. Herring become sexually mature after about 3 years, and when spawning, a single female herring can lay 20,000–40,000 eggs on the solid seabed over a few nights. Before the eggs have stuck to the seabed and each other, they have been fertilized by the males. Areas of up to 100,000 m^2 have been found on the seabed with four to eight layers of herring eggs, corresponding to several hundred billion eggs. The eggs hatch into larvae after a few weeks.

Herring roe have a fairly large water content of about 80%. The protein content varies quite a bit and is around 10–20%. The fat content is surprisingly low for a fatty fish, 2–4%, and the ω-3/ω-6 ratio varies between 10 and 20. The small fat content of herring roe is due to the fact that the eggs hatch after only 2.5 weeks and therefore require less nutrition than, for example, salmon eggs, which only hatch after up to 20 weeks. The finest quality of herring roe is obtained from fish that are unstressed and not pressed during capture and transport. Damaged roe are dark and discolored.

Herring roe have a crunchy mouthfeel similar to flying fish roe (*tobiko*, cf. Sect. 5.6), and colored herring roe are therefore used as a substitute for the more expensive *tobiko*. The smallest eggs are considered the best quality, which is achieved if the roe are not too ripe.

There is no exact data available for the annual world production of herring roe, but it is probably around 5000–6000 tons. The Japanese purchase the majority of the world's production of herring roe in the form of salted roe, *kazunoko*. The main body of loose, salted herring roe is produced from North Atlantic herring at the Danish factory Scandic Pelagic in Skagen, whereas whole, salted roe sacs are predominantly based on Canadian herring.

In Europe, it is only in coastal areas that herring roe have traditionally been eaten, often as a preparation alongside fried herring. Fried herring roe are a Dutch specialty. There is an Icelandic product, *stromluga*, which is made from smoked herring roe flavored with lemon and dyed black with squid ink. *Stromluga* has a strong smoky and slightly bitter taste and is suitable for a salad or pasta dish.

5.10.1 Salted Herring Roe

Whole salted roe sacs of herring roe with the outer membrane removed are a major specialty in Japan, where the salted roe are known as *kazunoko* (cf. Fig. 5.9). The Japanese term *kazunoko* means "many children" and is considered a symbol of fertility and a large family. This is also the reason why *kazunoko* appears on the special menu (*Osechi-ryōri*) at traditional Japanese New Year celebrations.

The roe are not removed from the freshly caught fish but only after it has been frozen, preferably in cold, salty water, which helps to preserve the shape of the roe sacs. If the two roe sacs are perfect and uniform in shape and size, *kazunoko* is given a particularly high value. The shape of the solid mass of roe, which is light yellow, resembles elongated crystals and is therefore called "yellow diamonds."

Salting herring roe for *kazunoko* typically takes 5–7 days. The process is not only intended to preserve the roe and make the eggs firmer but also to reduce any discoloration and traces of blood. For the same reasons, bleaching with hydrogen peroxide is sometimes used. Before eating the roe, it is desalted and typically marinated in *dashi* and possibly soy sauce. In Japanese cuisine, pieces of *kazunoko* are used as a topping on both *nigiri-sushi* and *donburi*.

Fig. 5.9 *Kazunoko*, salted herring roe

Loose, salted herring roe can be used to mix into a green salad or a seaweed salad to add a little color, umami, and an interesting crunchy mouthfeel. As with a wide range of other fish roe with small eggs, herring roe are also used in *taramasalata*.

5.10.2 Surströmming

There is a local strain of herring found in the Baltic Sea and the Gulf of Bothnia. This Baltic herring, also called *strömming*, is smaller than the common Atlantic herring, growing to just 15–20 cm. This small species of herring is probably best known, or perhaps infamous, for its use as *surströmming*.

Surströmming is made from herring that are caught in the spring, lightly salted, and fermented in barrels for up to 2 months, after which it is canned. Lactic acid fermentation then takes place in the cans, which develops the acidic substances that give the final product its name. At the same time, amino acids with umami taste are formed. Often, in cans with whole fish (without head), there will be well-preserved roe sacs. This sour and umami-rich roe are a special delicacy that can be eaten in the same way as *surströmming* with sour cream on slices of boiled potatoes.

On a Field Trip: Herring Roe—a Danish Specialty—in Japan
There are hardly many Danes who have tasted herring roe. In Japan, however, salted herring roe are a sought-after and expensive delicacy. The salting helped to give the roe firmness and a crunchy mouthfeel.

A Danish engineer Susane Røntved established a herring factory, Unifish A/S, in Skagen in 1981, and together with two other companies she developed an automatic system that could remove and repeatedly wash the roe during the filleting process so that the loose eggs could be released before freezing. She then developed a special salting technique so that the product could satisfy the quality requirements of the Japanese market. The Japanese customers then mastered a method of "gluing" the loose eggs together into blocks that were shaped like the herring's entire roe sacs but now without a roe sac membrane. For most of a 20-year period, Danish production of this form of herring roe dominated the world market.

In 2008, Unifish A/S was bought by Skagen Fiskeeksport, which after several mergers with other companies became Scandic Pelagic A/S, one of the world's leading companies in the pelagic fishing industry and specializing in herring products. Scandic Pelagic is located in Skagen at the very northern tip of Denmark and is owned by FF Skagen A/S. It is now the only Danish company that processes herring from the raw material. The factory, which is the world's largest herring factory with 200 employees, processes 100,000 tons of herring per year and approximately 1000 tons per day.

In January 2023, we had the opportunity to visit Scandic Pelagic in Skagen, where CEO Johannes Palsson and sales director Benjamin Bosse showed us around (Fig. 5.10). To our luck, 1000 tons of so-called Norwegian spring-spawning herring was unloaded from a giant cutter that day for filleting. In very cold water (−1 °C), the herring from the cutter is pumped into the factory, and the temperature is kept below 4 °C throughout the cooling chain until the final product. Unlike North Sea herring, which arrives in the fall, the roe from this strain of herring are not used directly, but we can still see during the visit how the herring are cut up on large filleting machines and the intestines and any roe are cut off. When the roe are used, it is washed repeatedly, and the eggs in the roe sac have already been loosened by the enzymes in the herring's stomach.

Scandic Pelagic exports loose herring roe to the Asian and North American markets, where it is used as kaviar. Some goes to the Eastern European market, where the roe are used in spreadable products. Recombined roe sacs from loose herring roe from the Danish herring factory are no longer produced.

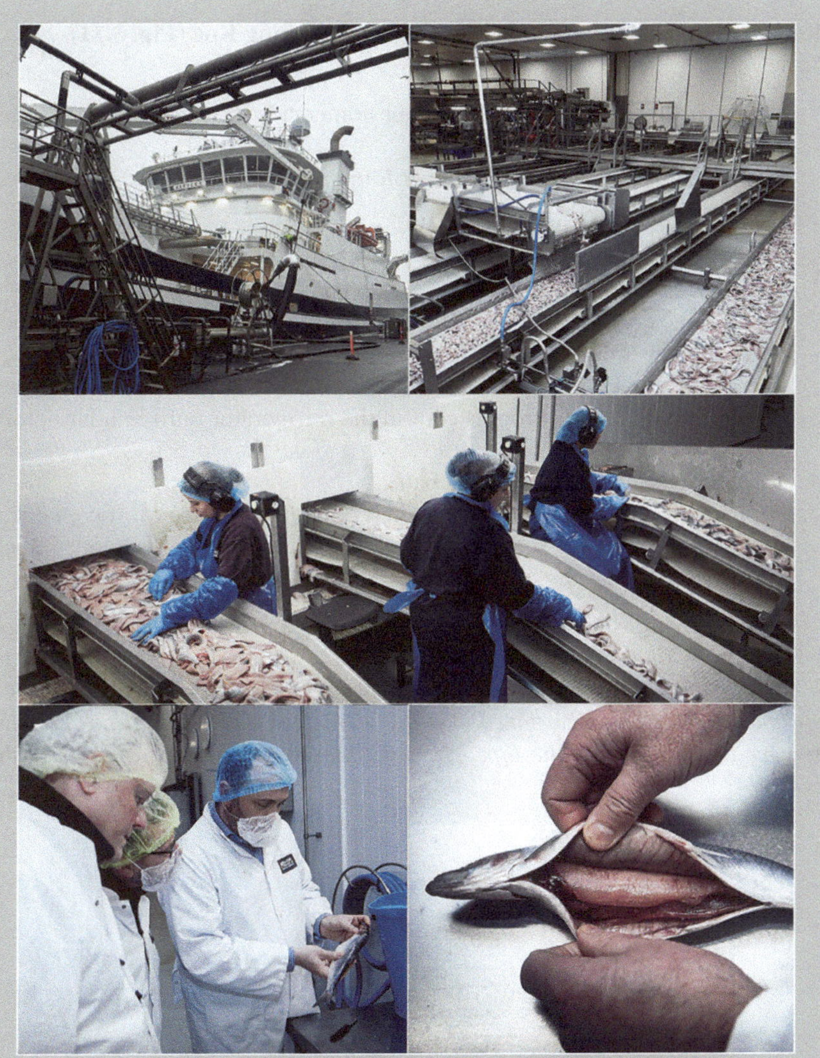

Fig. 5.10 Production of herring filets and herring roe at Scandic Pelagic facilities at Skagen, Denmark

Recipe 5.3 Poached *Shiitake* in *Dashi* with Herring Roe (Fig. 5.11)
Serves 4

8–10 g *dashi* powder and ½ l water or homemade *dashi**
50 g dried *shiitake* mushrooms
4 tbsp herring roe, flying fish roe, or *löjrom/sikrom*
½ tsp sesame seeds
Green *shiso* leaves (or chervil)

1. Soak dried *shiitake* mushrooms.
2. Cut the mushroom heads into slices and place them in ½ l *dashi* together with the mushroom sticks. Simmer the mushrooms over low heat for 10 min. Discard the sticks.
3. Let the broth and mushrooms cool in the refrigerator until ½ hour before serving.
4. Toast the sesame seeds on a dry pan.
5. Arrange the mushrooms and broth in a bowl, place a tbsp of roe on each, sprinkle some sesame seeds on top, and sprinkle with finely chopped fresh or dried *shiso* leaves.

**Homemade* dashi*: ½ l boiled cold water, 5 g* konbu *seaweed, 12 g* katsuobushi *flakes. Soak the* konbu *for 30 min in the water in a pot, then heat to 60 °C for 30 min. Remove the* konbu *seaweed, heat the water to 60–70 °C, sprinkle the* katsuobushi *flakes into the water, let them sink to the bottom, remove any foam from the top, strain the* katsuobushi *flakes, and the* dashi *is ready.*

Fig. 5.11 Poached *shiitake* in *dashi* with herring roe, *shiso*, and toasted sesame seeds

5.10.3 Herring Roe on Kelp

Herring spawn their roe on rocks and sandy bottoms but also on seaweed leaves of large kelps. Herring roe on *konbu* seaweed or other brown seaweed species with lamellar leaves is a sought-after and expensive delicacy in Japan, where it is called *kazunoko-konbu*. The finest and most valued products are those in which the roe sit in an even layer on both sides of the seaweed leaf.

Fig. 5.12 *Kazunoko-konbu*, herring roe on macrokelp (*Macrocystis pyrifera*)

In Alaska, 15% of the commercial production of herring roe comes from roe spawned on seaweed. The herring lays a layer of eggs on leaves of especially large seaweed species such as macrokelp (*Macrocystis pyrifera*) (cf. Fig. 5.12), and in some cases the layer can be up to 2 cm thick. The kelp leaves are a perfect substrate for the sticky roe, which can be spawned on one side but most often on both sides of the seaweed leaves. Although a niche product, commercial exploitation of herring roe in this way is more sustainable than the usual method, where the roe-laden herring are caught and the roe sacs are removed when slaughtering the fish.

The seaweed leaves with the roe are usually cut into strips and put in a brine or frozen. The strips can be eaten as a snack or as a starter after being thoroughly soaked in water. The slightly tough, yet succulent texture of the seaweed goes well with the crunchy crispness of the roe and the salty, slightly bitter taste of the fish roe and brine.

Since *kazunoko-konbu* is a rare and exclusive product, there have been attempts to make imitations by pressing loose herring roe onto pieces of seaweed, where they are glued together using the same gelling agents that are also used to make *surimi* (minced and glued together meat from fish and shellfish).

The indigenous peoples of the northwest coast of Canada have traditionally used herring roe on seaweed as food, and the roe were a prized part of a feast, as well as an important trade item. Large seaweed leaves were harvested and anchored under rocks on the seabed off an estuary to increase the chances of herring spawning on the seaweed. The roe on the seaweed leaves were preserved by drying on rocks in the sun, after which they could be bundled and transported for trade. Cut into strips, the dried herring roe on kelp were used as a snack or candy. The dried roe could also be easily resoaked in water or boiled and fried with fat.

5.11 Ling

Ling (*Molva molva*) belongs to the cod family. Its roe are not commonly consumed, but it can in principle be used like roe from other codfish. Dried ling roe in brine (*huevas de maruca*) have been a specialty in Andalusia since ancient times.

5.12 Lumpfish

Kaviar from lumpfish (*Cyclopterus lumpus*) is nowadays a popular and gastronomically valued roe product, which, depending on the market, used to be relatively inexpensive in relation to the quality. Lumpfish are caught off the coasts of Norway, Canada, the USA, and especially Greenland and Iceland, which together account for over 95% of the global production of lumpfish roe. Sweden, Canada, and Iceland are world leaders in the sale of roe products from lumpfish. A large lumpfish female can carry 100,000–200,000 eggs, equivalent to 700 g of roe. The annual global production of lumpfish roe varies from year to year but is typically 3000–4000 tons. The catch has been decreasing in recent years, and it is recommended to exploit lumpfish more sparingly.

The roe are usually extracted from the fish on the fishing vessels, and most of it is salted in barrels on the cutter. Sugar is also added for some markets. Lumpfish roe contain approximately 80% water, 10–15% protein, and 4–6% fat. The ω-3/ω-6 ratio of around 30 is extremely high.

The first lumpfish roe products began to appear on the market in the 1930s as a substitute for real caviar. Lumpfish eggs are approximately the same size as sevruga eggs, around 2–3 mm, and were dyed black as a substitute to resemble sturgeon caviar. At that time, lumpfish roe were not particularly sought after or exploited, which may be due to the fact that the roe have a very neutral and not particularly strong or distinctive taste. It is a synthetic tar dye (brilliant black) that is now used to dye lumpfish roe black.

Commercially produced lumpfish kaviar, which is sold in glass jars, is lightly salted, has preservatives added, and must be stored in the refrigerator. The uncolored roe, which are only sold in small quantities, are light white and slightly reddish. Some lumpfish roe are pasteurized and tolerate this heat treatment better than other types of roe. On the European market, one third of the lumpfish roe are sold as black and marketed as a caviar substitute. Some lumpfish roe are also dyed with yellow, green, and orange dyes. However, it is mostly the red and black versions that dominate the market (cf. Fig. 5.13). For all colored products, the disadvantage is that the colors can transfer to other foods, e.g., an egg dish.

Apart from the large-scale industrial production of preserved roe products from lumpfish, fresh and untreated lumpfish caviar in season are a sought-after food in traditional ways as kaviar, in dressings, and as an accompaniment to and garnish for, e.g., egg and fish dishes. Lumpfish roe are one of the only types of fish roe that are often eaten completely raw and without salting.

Fig. 5.13 Lumpfish roe in three different colors, natural, black, and red

The loose eggs can be released from the roe sac by mechanical whipping in a 3–5% salt solution, and some industrial processes use mechanical separation from the connective tissue of the roe sac by pressing the roe through special sieves. During the subsequent treatment with salt, the eggs initially stick together, but gradually the salt will draw liquid out of the eggs by osmosis and detach them from each other. However, if this salting process is allowed to continue for too long, the eggs clump together.

A traditional use of fresh lumpfish roe is as a starter or canapé, garnish on fish dishes and sandwiches, or in egg dishes with halved or sliced hard-boiled eggs. Lumpfish roe are a very popular dish especially in season and were previously reasonably priced but have become quite expensive in recent years.

5.13 Mackerel

There are a number of species of mackerel in different parts of the world, e.g., *Scomber japonicus, Scomber australasicus*, and *Scomber scombrus*. The eggs are small, 0.8–1.0 mm. Mackerel roe are not a common product. In India, the roe are considered a delicacy, which is deep-fried and served with chili paste. In Portugal, whole mackerel roe sacs are canned with olive oil.

5.14 Monkfish

Monkfish (*Lophius piscatorius*) carry eggs with an egg diameter of 2–3 mm. The eggs are oval and orange. However, it is rare for ordinary consumers to have access to these eggs.

5.15 Mullet

There are different species of mullet, e.g., gray thick-lipped mullet (*Chelon labrosus*), but the most well known in connection with roe is the flat-headed mullet (*Mugil cephalus*). The eggs are small, around 0.6 mm. Mullet roe are particularly well known in classical and ancient gastronomy because they were the most valued type of roe for producing dried and pressed whole roe sacs (*bottarga*), which we will discuss elsewhere in the book (cf. Sect. 11.4.1). Mullet roe have a particularly high content of waxy fats, which can constitute 60–70% of the oil content and which gives the dried product a special texture like wax. Fresh mullet roe contains 50–60% water, around 25% protein, and 15–20% fat. The cholesterol content is quite high, around 0.5%.

5.16 Pollock

It is especially the roe from Alaskan pollock (*Gadus chalcogrammus*; formerly *Theragra chalcogramma*) that are eaten in many parts of the world. Alaska pollock is a codfish, and since many of the products from its roe are the same as those made from cod roe, they are discussed under Cod in Sect. 5.4. Alaska pollock are one of the largest catches in global fisheries, and most of the roe ends up on the Japanese market.

The roe typically makes up 5% of the fish's weight. The eggs are 1.3–1.5 mm in diameter and contain 20–25% protein, about 4% fat, and 0.3% cholesterol. The ω-3/ω-6 ratio is relatively high (about 15).

The roe of Alaska pollock can be processed into the salted and seasoned roe products *mentaiko* and *tarako*, as described under Cod in Sect. 5.4.

Two other species related to Alaska pollock are also called pollock, *Pollachius virens* and *Pollachius pollachius*. Their roe can be used in the same way as for Alaska pollock, but roe from these species are not a major commercial product.

5.17 Salmon

Salmon produces large orange-red eggs, typically 4–8 mm, significantly larger than sturgeon eggs and slightly larger than trout roe. The eggs are shiny and partly transparent. Kaviar from salmon roe was first developed in Russia in the 1830s as a substitute for caviar and has often been called "red caviar" (*ikra*). Russia is probably associated mostly with sturgeon roe, but in fact the Russian market consumes around 15,000 tons of salmon roe per year. This red kaviar is also, like caviar, produced a la *malossol*, i.e., with a little salt, around 3%.

Fig. 5.14 Red kaviar from salmon roe

There are a wide variety of different species from the salmon family (Salmonidae) from which the roe are used, e.g., *Oncorhynchus keta* (chum salmon), *Oncorhynchus kisutch* (coho), *Oncorhynchus nerka* (sockeye), *Oncorhynchus gorbuscha* (pink or humpback salmon), *Oncorhynchus tshawytscha* (chinook), and *Salmo salar* (Atlantic salmon).

Salmon roe have a low water content, 45–55%, and a correspondingly high content of protein (20–30%) and fat (10–20%). The cholesterol content is quite high and typically around 0.5%. Salmon roe have some of the highest values of the ω-3/ω-6 ratio (about 10–20) for roe. The content of triglycerides is high and comparable to the content of phospholipids.

On the world market, it is mainly Pacific salmon roe that are fished by Japan, the USA, Canada, and Russia, and it mainly concerns chum salmon (*Oncorhynchus keta*) and pink salmon (*Oncorhynchus gorbuscha*) but also a few other species. These wild species are caught in nets. Atlantic salmon (*Salmon salar*) from aquaculture is also used as caviar, and here there is only a small market from wild populations. Another species in the salmon family is arctic char (*Salvelinus alpinus*), from which roe products are also produced.

It is particularly roe from salmon aquaculture in Norway and Finland that contribute to the world market, and the quality is high. Previous problems with an "aquaculture off-flavor" have now been overcome by adjusting the fish's feed.

A significant part of salmon roe ends up on the Japanese market, where salmon roe are called *ikura* in the form of loose eggs, e.g., chum-*ikura* (cf. Fig. 5.14). In Europe, the largest market is in Germany, followed by France. Whole salted roe sacs, *sujiko*, are also used in Japan. The Japanese interest in salmon roe is of recent date, and it was the Russians who introduced the method of salting salmon roe to the Japanese market in 1912. Here, however, producers were quick to add more characteristic Japanese flavors to the roe from *dashi*, soy sauce, and sake.

The specialty product *sujiko* is made from whole roe sacs that are first wet salted for 15–20 minutes in a salt solution, which may also contain added flavorings, stabilizers, and preservatives, such as nitrites and polyphosphates. The roe sacs are then sprinkled with fine salt and placed under pressure for 3–5 days at temperatures below 11 °C. Sockeye salmon (*Oncorhynchus nerka*) is particularly used for this. *Sujiko* is a traditional dish in Japan at the annual Obon festival in August, when deceased ancestors are celebrated throughout the country. A by-product of *sujiko* production is *barako*, which consists of the loose eggs that may break free during the process. *Barako* is used in sushi and as a filling in rice balls (*onigiri*).

In the food culture of the indigenous peoples of the American northwest coast and in Alaska, there is a tradition of making compressed and dried roe sacs wrapped tightly in skins, whereby the roe are fermented into a kind of solid "cheese" of salmon eggs, so-called stinking eggs.

The main production of loose salmon eggs as red kaviar traditionally takes place by releasing the eggs from the roe sac, which is first cleaned with a 3% salt solution, by mechanically pressing the roe through sieves. Modern techniques use an enzyme, collagenase, which dissolves the collagen fibers that bind the eggs together. The process corresponds to the biological process that takes place in the female fish before spawning. The loose eggs are then treated in a salt solution for flavor development and preservation. This process hardens the outer shell of the eggs, and the resulting raw product is *ikura*. Some *ikura* is then marinated, depending on the market for which the red kaviar is intended. For the Japanese market, this involves marinating in, for example, sake, soy sauce, sweet rice vinegar, or other marinades, possibly with spices.

In Europe and North America, *ikura*, which is called red kaviar there, is used as a starter, in sauces for pasta dishes, in egg dishes, and as a garnish or accompaniment to, for example, fish dishes. *Ikura* is used in both hot and cold dishes. Often, the roe that are served as salmon roe in these contexts are, however, the cheaper trout roe.

The taste of salmon roe is generally less "fishy" than that of caviar. This is mainly because, unlike caviar, salmon roe are not usually matured, and the polyunsaturated fats are therefore less exposed to oxidation.

5.18 Sardine

The sardine family includes a number of species, such as the European sardine (*Sardina pilchardus*), the Japanese sardine (*Sardinops melanostictus*), and the Californian sardine (*Sardinops sagax caerulea*). The eggs are quite small, 1.1–1.3 mm. The roe can be eaten together with the fish meat. In some places, the roe are salted and smoked. Sardine roe are sold in oil in Portugal. In India and Pakistan, sardine roe are a delicacy when fried or in a kind of curry (*gashi*).

5.19 Smelt

The smelt family (Osmeridae) includes a number of smaller species, e.g., capelin (*Mallotus villosus*) (see Sect. 5.2). The roe from these fish are used in the form of the product *masago*.

5.20 Sprat

Sprat (*Sprattus sprattus*) is a small, pelagic schooling fish. It is found and fished in very large quantities in European coastal waters, in particular in the North Sea, the Baltic Sea, and the Mediterranean. Sprat seldom reaches the consumer as fresh fish. Most of the catch goes to animal feed, which is a shame, as it is a fine consumer fish. The roe resemble ordinary herring roe, and the eggs have a diameter of 0.8–1.5 mm.

5.21 Sturgeon

The different species of sturgeon and their roe were described in a separate chapter, Chap. 3. Depending on the species, sturgeon eggs contain 50–60% water, 20–25% protein, and around 15% fat. Sturgeon roe are therefore as fatty as salmon roe. The cholesterol content is high, around 0.6%, and thus somewhat higher than the typical content in salmon and trout roe. Depending on the species, the egg size varies from 2 mm to 4 mm.

5.22 Trout

Trout belong to the salmon family (Salmonidae) and include a number of species, such as rainbow trout (*Oncorhynchus mykiss*), which lives in fresh or low-salinity water, and the ocean-migrating sea trout or brown trout (*Salmo trutta*), of which there are a number of varieties. Trout produce large red eggs, approximately 3.5–4.5 mm in diameter, i.e., larger than caviar, but not as large as salmon, and their color is slightly more orange but just as shiny. The whole ovaries of trout are often used as *sujiko* (cf. Fig. 5.15).

The source of commercial trout kaviar is mainly rainbow trout, which is raised in fish farms, but it is also produced from wild stocks, called salmon trout in Europe, which behave like salmon. Farmed trout produce eggs that are more golden because

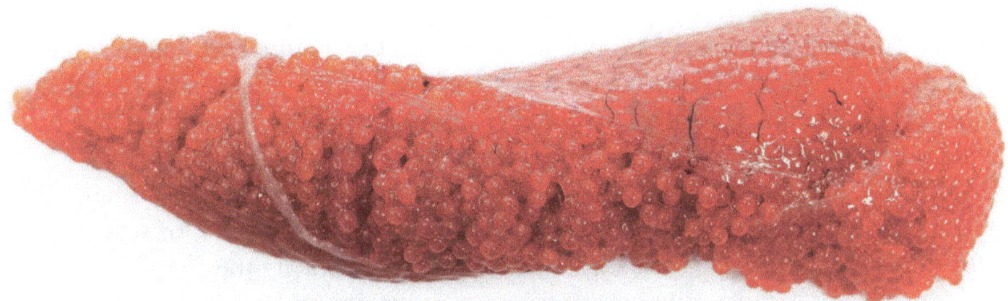

Fig. 5.15 Whole rainbow trout ovary used for *sujiko*

they often lack the red pigment astaxanthin, which can, however, be supplied through their feed. The body weight of a spawning trout can consist of 10–20% roe. Trout roe are processed like salmon roe into kaviar products with loose eggs and have generally the same types of uses in the kitchen, e.g., lightly salted as *ikura*. Trout roe have a taste and texture slightly different from salmon roe, which reflects the species difference and farming conditions.

Like salmon roe, trout roe have a low water content (around 60%) and a correspondingly high content of protein (around 25%) and fat (around 10%). The cholesterol content is lower than in salmon roe. The ω-3/ω-6 ratio is not particularly high, around 2, and thus lower than in salmon roe and most other types of fish roe.

On a Field Trip: Danish Production of *Sujiko* for the Japanese Market

The Priess family is a Danish family dynasty, which has influenced the Danish fishing industry for over 100 years. It started in 1900, when the 14-year-old Anders Priess started his own company with local fish trading in the small town Glyngøre. The company was called Priess and Co, which covered up the fact that Anders' mother had to stand as the formal owner due to Anders' young age. The young man's fish trading was not well received by Glyngøre's older fishmongers, who did not have much to spare for such a greenhorn.

Little could they have foreseen that the company now over 120 years later has grown and thrives with the fourth-generation Priess at the helm. It is a remarkable piece of Danish industrial history, but the main reason that AquaPri, as the company is now called, should be included in this book is that it produces trout roe of such a quality that the roe have found favor with the otherwise very picky Japanese market. AquaPri produces three types of trout roe products, two of which are made from immature "green eggs," which are taken from the fish in November to January. Every year in November, a team of Japanese technicians comes to Aarøsund, where AquaPri's roe factory is located, and processes, quality checks, and packages the whole ovaries of "green eggs," which are turned into *sujiko* for sale on the Japanese market, especially in Hokkaidō. Other "green eggs" are also used by AquaPri all year round to produce loose eggs, *ikura*, which have a relatively soft shell. The product from the soft *ikura* is sold frozen and used in the catering sector. The third of AquaPri's trout roe products is based on the mature eggs harvested in February to March. The product is an *ikura* with firmer eggs and a thicker shell. It is an exclusive and more expensive product than the soft-shelled eggs, and it is sold to the consumer in glass jars.

In 2012, cousins Henning and Morten Priess took over the management of AquaPri, which now employs over 100 employees in Aarøsund, of which 25 are Japanese technicians who work at the company during the *sujiko* season. There are only a few permanent employees at the company, which during the season is based on a predominantly Eastern European workforce who works around the clock in three shifts. These workers are responsible for transporting the live fish from boat to quay, killing, cutting up, removing the egg sacs, and cleaning the body of the slaughtered fish. Japanese technicians then take over the sorting of the roe (according to color, maturity, and size) as well as the salting and packaging of the finished *sujiko* for the Japanese market.

The fish for roe production are raised on AquaPri's own mariculture farms around Denmark. AquaPri, which is Europe's largest producer of trout roe, turns around 500 tons of trout roe per year, depending on demand on the international market, of which around a fifth goes to the Japanese market.

The trout have an average weight of 3.7 kg and are around 3 years old when they are slaughtered and the roe are removed. On average, the roe constitute around 10% of the trout's slaughter weight. These farmed trout are less fatty than Norwegian salmon, and a third of their feed is from vegetable sources. Due to the fish's good nutritional status, the fish meat can also be used as a food product. This is in contrast to wild trout and salmon, which are often completely emaciated when their ovaries are fully developed.

The crucial thing for producing the highest quality *sujiko* is that there is a very short time between the fish being killed, the roe being removed, and the time they are preserved in salt.

In November 2022, one of the authors (Ole) had the opportunity to visit AquaPri's factory in Aarøsund, where the Danish production of *sujiko* takes place. Kurt Mogensen showed us around.

The live trout are sailed from the fish farm to the port of Aarøsund in so-called well boats, from where they are led through a pipe with flowing water and an electric field in a facility on the quay, which stuns the fish. They are then quickly slaughtered with a stab through the gills. They are then bled in cold water while being transported the last 100 meters into the factory to a mechanized line operated by about ten people. The fish are automatically placed head up and belly forward in mesh on a conveyor belt, where they are then processed by hand. First, the belly is cut open, and the roe sacs are carefully removed, after which the trout continue on the line to other employees who remove the entrails and clean the open belly. After further cleaning with clean water, the fish carcasses are taken to a sorting facility and then to freezing.

The roe are taken from a conveyor belt onto a sorting table (see Fig. 2.8), where the roe sacs are sorted by hand according to size, color, and degree of maturity. A Japanese team of specialists from the large Japanese fish products company Mitsui & Co is in charge of this work, and on the day of the visit the team is led by Keiichiro Hayakawa, who supervises the quality control.

Hayakawa-san explains the process of making *sujiko* quite openly. As he says, it is quite simple and follows classic recipes: washing, salting, pressing, and ripening the entire ovary before freezing. But the prerequisite is that the roe are of top quality and have no "off-flavor." It must not smell of fish, and it is also remarkable that there are no marine aromas to be detected anywhere in the cold and meticulously cleaned factory.

The roe of trout slaughtered in November to January consist of "green" and immature eggs, and the membranes between the eggs and the egg sac are intact. The eggs in the ovary are connected like a bunch of grapes (cf. Fig. 5.15). The sorted roe sacs are washed briefly in a dilute chlorine solution for disinfection, after which they are wet salted for 15 minutes in a saturated salt solution (26% NaCl) together with small amounts of sodium nitrite. Sodium nitrite, which enhances the red color of the roe, is not permitted in Danish food but may be used in Denmark in the production of food for export. The nitrite content in the product is below 5 ppm, which is permitted in Japan.

The Japanese technicians now carefully place the roe sacs in layers in open special plastic boxes with space for approximately 5 kg (Fig. 5.16). Fine salt is sieved between the layers so that all surfaces are covered. This salting, as well as the previous wet salting, helps to make the individual eggs firmer. During packaging, there is a top on the open boxes corresponding to approximately 10% of the weight. The boxes are now placed in layers with plates between them and pressed together for 2 days, causing about 10% of the weight to drain in the form of liquid. The boxes can then be tightly lidded.

The packaged *sujiko* is stored for a week at about 10–15 °C. During this period, the roe develop more flavor due to enzymatic activity, which releases umami taste. The product is then frozen before being exported to Japan, especially for sale on Hokkaidō, the northernmost island in Japan, where people traditionally eat salmon. Hayakawa-san explains that there is still a good market for *sujiko* in Japan, but the market has been declining in recent years.

The team of Japanese technicians travels around the world, piecing together seasons for the production of *sujiko* with roe from pink salmon (*Oncorhynchus gorbuscha*) in Alaska from mid-June to late August, roe from chum salmon (*Oncorhynchus keta*) in Hokkaidō from early September to mid-October, and then the Danish season with rainbow trout (*Oncorhynchus mykiss*) from late November to late December in Aarøsund.

The part of the trout roe that is not used for *sujiko* in Aarøsund is frozen whole and used out of season to produce loose roe, *ikura*. After thawing, the roe for *ikura* are salted as usual and then treated with enzymes that dissolve the membranes. The loose eggs are frozen again in 50 g to 10 kg packages, which are then sent to producers who repackage them for retail, where the consumer can encounter Danish trout roe in glass.

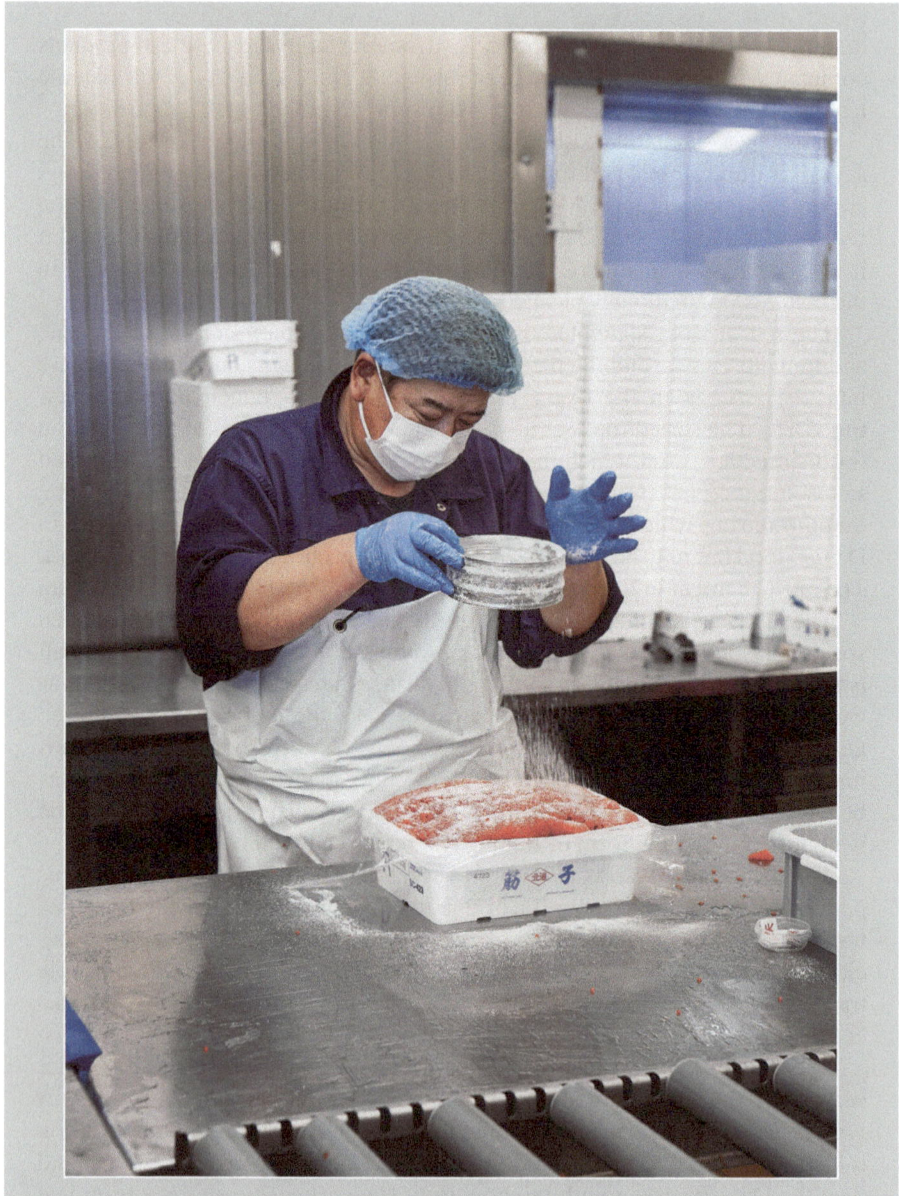

Fig. 5.16 Preparation of trout roe for *sujiko*

5.23 Vendace and Whitefish

White fish (*Coregonus lavaretus, Coregonus clupeaformis, Coregonus maraena*) are related to the smaller species vendace (*Coregonus albula*). Both species belong to the salmon family (Salmonidae) and are both somewhat confusingly called whitefish. Whitefish is a brackish water fish that is common in the Gulf of Bothnia.

The roe from the vendace is sought after and is marketed in Sweden under the name *lögrom*. The Swedes are large consumers of *lögrom*, but most of it is actually imported frozen from North America. Roe from whitefish are called *sikrom*. The eggs from both species are small, 0.9–1.4 mm, and pink (cf. Fig. 5.17). The roe are often used raw or only slightly salted in the same way as caviar and as a garnish. It has a slightly smoky taste. A traditional product is the Greek *taramasalata*, which is a mass of bread or boiled potatoes mixed with roe (which can also be from cod, mullet, or carp), olive oil and lemon juice, possibly pepper, and spring onions. Roe from the whitefish and vendace, like roe from other salmonids, have quite a high amount of fat, around 10%, and the triglyceride content is relatively high.

Kalix *Lögrom*: A Protected Food Designation

Kalix is a small Swedish town in the archipelago at the bottom of the Gulf of Bothnia. Here, people traditionally harvest roe from vendace (*Coregonus albula*) from the end of September until the spawning season in late October. The roe are more orange red than *lögrom* from other locations, and its taste is milder and determined by the influx of fresh water into the bay. The composition of the roe is therefore also unusual, especially with regard to the ratio between barium and strontium. In recognition of the gastronomic quality of this roe, Kalix lögrom was granted special status with a protected food designation (PDO—protected designation of origin) by the European Union in 2010. This roe is always served at the banquet when the Nobel Prizes are awarded.

Fig. 5.17 Roe from the whitefish (*sikrom*) on the left and the vendace (*lögrom*) on the right

5.24 Wolffish

Seawolf (Atlantic catfish, *Anarhichas lupus*) have eggs with an unusually large diameter (5–7 mm) for a saltwater fish, actually comparable to trout. The eggs are dull yellow. The eggs are not a commercial product and do not normally reach consumers.

5.25 Zander

Zander (*Sander lucioperca*) is a common freshwater fish in the perch family. There is only scant knowledge about the ingredients in zander roe. The eggs are 2.0–2.5 mm. You may be lucky enough to get a zander with roe, and the roe are suitable for salting or boiling. Roe from zander are also known as *galagan* and are not a common commodity.

Recipe 5.4 Poached Terrine of Various Types of Roe in *Nori* Seaweed (Fig. 5.18)
Serves 4

300 g of various types of roe, e.g., smaller types of flatfish roe sacs
1 dl rice wine vinegar
½ dl soy sauce or *ponzu*
1 piece of *nori* seaweed
Pickled ginger
Salt

1. Clean the various types of roe and remove the coarsest membranes.
2. Cut the large roe sacs into bite-sized, smaller pieces; cut where it seems natural to divide.
3. Place the roe in a bowl and marinate it with rice wine vinegar and soy sauce.
4. Place the bowl in the refrigerator for 2–3 hours.
5. Place a piece of *nori* seaweed on the kitchen counter and lightly moisten it with a few splashes of the marinade.
6. Remove the roe, save the marinade, place the roe across the first third of the *nori* seaweed, roll it tightly into a sausage, and place it on cling film, which is also rolled up and tightened tightly. Tie the ends together.
7. Find a pot that is wide enough to contain the roll.
8. Pour the marinade, salt, and enough water to cover the roll into the pot and simmer it under a lid over low heat for about 20 min, depending on the thickness.
9. Let the roll soak in the water for 20 min, and refrigerate the roll until ready to use.

Fig. 5.18 Poached terrine of various types of roe in *nori* seaweed

The terrine can be served with pickled ginger, *wasabi*, or *miso*-mayo. For this dish, it is a good idea to collect different types of roe over time and store them in the freezer until needed.

Recipe 5.5 Raw Marinated Roe with Mustard and Tomato Seeds (Fig. 5.19)
Serves 4

250 g roe,* if possibly as whole roe sacs, preferably from different fish
4 tablespoons red wine vinegar
½ tablespoon blanched mustard seeds
½ teaspoon sugar

Fig. 5.19 Raw marinated roe with mustard and tomato seeds

1 tablespoon ketchup
12 small ripe tomatoes in different colors
2 spring onions
2 stalks broad-leaf parsley
1–2 lemons
4 tablespoons good olive oil
A little whole horseradish
Freshly ground pepper and salt

1. Remove the membrane from the semi-frozen roe sac, cut it into thin slices, and place the slices on a serving platter or four plates.
2. Blanch the mustard seeds for 1 minute in boiling water.
3. Mix the red wine vinegar, mustard seeds, sugar, and ketchup, and let the marinade sit in the fridge for a while.
4. Cut the tomatoes in half, carefully remove the seeds, and save them. Thinly slice the spring onions and cut the parsley into strips.
5. Spread the tomatoes, tomato seeds, spring onions, and parsley over the roe.

6. Grate some lemon zest over the roe, squeeze the lemon juice on, spread the marinade, and drizzle with a little olive oil.
7. Grate some horseradish to taste.

The raw marinated roe are a kind of ceviche. It has a strong umami synergy from the combination of roe with tomato. The texture becomes interesting when combined with fish eggs, mustard seeds, and tomato seeds.

**Roe can be collected in the freezer until you have enough for one serving. In any case, the fresh roe you use must have been frozen for at least 24 hours.*

Recipe 5.6 Pasta *Tarako* (Fig. 5.20)
Serves 4

100 g roe, fresh or in a jar, e.g. salmon or lumpfish roe
250 g spaghetti + a little cooking water
2 tbsp soy sauce
2 dl cream
2 tbsp melted butter
2–3 tbsp lemon juice
1 spring onion or green *shiso* leaves

1. Squeeze or scrape the roe out of the roe sac if using fresh roe.
2. Mix the roe with soy sauce, cream, melted butter, and lemon juice in a bowl.
3. Cook the spaghetti in lightly salted water in a large pot according to the instructions on the package, minus 1 minute. Pour off the water but save some.
4. Finely chop the spring onion or *shiso* leaves.
5. Pour the cooked pasta into a suitable pan with a few spoonful of boiling water and heat.
6. Add the roe mixture, stir until the spaghetti and roe cream are well combined, and the dish appears creamy.
7. Sprinkle with spring onions or *shiso* and serve immediately.

In this dish, the neutral pasta comes into its own as a recipient of the essence of all the flavors of the dish and with a pleasant bite.

Fig. 5.20 Pasta *tarako* with *shiso*

Recipe 5.7 Squid Fettuccine with Roe and Lobster (Fig. 5.21)
Serves 4

2 fresh lobster tails
Brine: 1 l water and 50 g sea salt
200 g whole piece of squid mantle
8–10 small very ripe sherry tomatoes
2 unpeeled limes
4 tbsp *ponzu*
4 tbsp olive oil for marinade and a little for frying
50 g salmon roe or other roe
50 g lemon Cavi-art pearls, or a little finger limes
1 dried black lime (*loomi*)

Fig. 5.21 Squid fettuccine with lobster, trout roe, and Cavi-art pearls with lemon juice

1. Cool the brine to 5 °C.
2. Cut the lobster tails in half and remove the shell and guts.
3. Put a wooden stick lengthwise through the lobster meat and place the 4 half tails in the brine for 5 min.
4. Freeze the squid, place it with the outside facing down, and cut it into long, even, thin strips for the fettuccine. Place the strips in a bowl.
5. Chop the tomatoes, press the tomato flesh and juice through a sieve, grate the zest of the green limes, squeeze the juice into them, and add the *ponzu* and olive oil.
6. Pour the marinade over the fettuccine and leave to cool for at least 15 min.
7. Quickly fry the lobster tails in olive oil on a hot pan.
8. Hold the end of the wooden stick and wrap the fettuccine around the lobster tails; carefully remove the sticks.
9. Arrange a tail with salmon roe and lemon Cavi-art pearls on top and pour the remaining marinade from the fettuccine over it.
10. Grate dried black lime zest (*loomi*) over it and serve.

Chapter 6
Roe from Crustaceans

Crustaceans (Crustacea) are a large suborder of ten-legged arthropods with several families, whose meat we eat and, in some cases, also the roe. This applies, for example, to lobsters, Norway lobsters, crabs, crayfish, and shrimps, which can be divided into five large families, crayfish, true lobsters (Nephropidae, e.g., lobsters and Norway lobsters), crabs (Brachyura, e.g., brown and littoral crabs), true shrimp, Caridea (e.g., rockpool shrimp and deep-water shrimps), and prawn (Dendrobranchiata, e.g., tiger prawn). Somewhat confusingly, we call some crustaceans for crabs; for example, the large Kamchatka crab (king crab) is not a crab but belongs to a completely different family of so-called intermediate crayfish (Anomura).

The reproductive organs of crustaceans are located under the carapace, and this is where the roe are formed. The roe can have different shades of red, orange, green, and dark green to completely black, and the color depends on which pigments are predominant. When the roe are immature in the gonads, they have the character of a jellylike mass, and many confuse them with the liver (hepatopancreas), which in these animals is actually one combined organ for the functions of the liver and pancreas. In small crustaceans such as shrimp, where the entire head is usually removed and only the tail is eaten, the immature roe are often not noticed, but in large crustaceans such as lobsters, it is clearly visible under the carapace as a dark green and almost black mass, which is considered a delicacy by many. When heated, the color of the roe changes to red, and if a lobster has not been cooked for a sufficiently long time, it is immediately visible by the fact that there is still some dark color left. The mechanism of the color change is the same that causes the color of crustacean shells to change to red when cooked.

O. G. Mouritsen, K. Styrbæk, *Roe and Roe Gastronomy*,
https://doi.org/10.1007/978-3-032-13142-3_6

Crustaceans May Change Color When Cooked

The carapace of living crustaceans such as lobsters, crayfish, and shrimps has gray-brown or gray-green colors, which are due to a molecular complex (crustacyanin) of a carotenoid (astaxanthin) and a protein. Astaxanthin, like many other carotenoids (e.g., carotene in carrots), is itself reddish orange, but as long as astaxanthin is bound in the complex with the protein, its reddish-orange color is not visible, but it is released and oxidized to astacene, which gives the lobster its reddish-orange color when cooked. It is predominantly the substance astaxanthin that is found in crustaceans, but in some species other carotenoids occur, e.g., canthaxanthin, which gives a more greenish color, e.g., in certain shrimps and crabs. In some cases, a species has a mixture of different carotenoids. Other carotenoids include doradexanthin, zeaxanthin, idoxanthin, and tetraol. Similarly, in lobster roe, there is another protein complex (ovoverdin), to which astaxanthin is also bound, hiding its red color. Ovoverdin is dark green and almost black, but when cooked the astaxanthin is released and the roe turns red.

Unlike fish, the eggs of crustaceans are not fertilized in the water but inside the female's gonads. During mating, the female stores the male's sperm for use at the right time, and the eggs are only released after they have been fertilized. When the eggs are mature in the ovaries and have been fertilized, they are in some species released directly into the water, and in others, such as lobsters, crayfish, and shrimp, the eggs are released and placed under the tail between the small swimlets, where they stick together in clumps. Here they remain until they hatch. The eggs of crustaceans have very different colors depending on the species and the predominant pigments. When crustaceans hold their eggs in this way between the swimlets, they are said to have "berries."

There are only very few studies of the ingredients, including nutrients and flavors, in crustacean roe, and it is mainly the immature gonads that have been studied. It would be expected that the composition of the roe of different species of crustaceans is broadly comparable. If data are available in the literature for specific species, it will be described below. Typical values are 60–75% water, about 20% protein, and 5–10% fat. As with fish roe, there is a large predominance of polyunsaturated fats, and the ω-3/ω-6 ratio is high (about 5). The cholesterol content is generally low.

Crustaceans are very delicate and go bad quickly after they die. They are therefore sold either live or cooked and frozen. The perishability applies even more to their mature roe, which decompose quickly and can discolor the water in which the crustaceans are stored. Preservatives have been added to some industrial products to prevent this from happening.

6.1 Crab

There are almost 5000 different species of crabs. Two examples include the small littoral crab (*Carcinus maenas*) and the slightly larger brown crab (*Cancer pagurus*). Crabs are best known in the kitchen for the fine, white muscle meat in their claws or as a flavoring for a crab bisque. There are also some who appreciate the brown meat that sits under the crab's carapace, which has a somewhat stronger and more complex sea flavor.

Crabs have a very small flap that is folded up under the tail. This is where the eggs are stored (cf. Fig. 6.1) for a few weeks after fertilization before being released and hatching, and the small larvae become part of the ocean plankton until they fall to the seabed. A female crab can produce over 100,000 eggs, depending on its species and size. The roe of the littoral crab is green or orange red, while the roe of the brown crab is orange red. The eggs are very small and typically less than 1 mm. Crab roe are rarely eaten as is, but they are often used as a flavoring and texture element in, e.g., soups.

Fig. 6.1 European spider crab (*Maja squinado*) with roe under its black flap

As previously mentioned, the Kamchatka crab (*Paralithodes camtschaticus*) or king crab is not a true crab. We mention it briefly here anyway, because its roe have begun to appear in some gourmet restaurants. The king crab can grow very large, with a leg span of up to 2 meters. The king crab is found in the Bering Sea and the Arctic Ocean. A female king crab can produce a large amount of roe. However, widespread commercialization of the roe is not expected for the time being. This is due both to the lack of the necessary technology to collect and process the roe and to the fact that the protection periods for the female crabs limit the possibilities. The roe consist of rather small eggs like herring roe, and the taste is somewhat neutral. On the other hand, the color is a beautiful reddish violet, and the eggs are shiny and glossy, and they have a crunchy mouthfeel. They are therefore great as a garnish for fish dishes, for example.

There are only a few studies of the ingredients in crab roe. An example is the brown crab (*Cancer pagurus*), whose roe contain 57% water, 25% protein, and about 3% fat. The cholesterol content is 0.17% and the ω-3/ω-6 ratio is about 4.

6.2 Crayfish

Crayfish typically live in fresh water, and in Europe it is the broad-fingered crayfish (*Astacus astacus*) that is of particular interest, but due to diseases in this species, it is the signal crayfish (*Pacifastacus leniusculus*) that most often reaches our dinner tables.

Like lobsters and shrimps, the immature roe (coral) are found under the carapace and are dark green to black in color, but this depends on the species. Here too, the color changes to red and orange when cooked. When the eggs are fertilized, they are secreted between the legs and stick together in clumps between the small swimlets on the underside of the hind body (cf. Fig. 6.2). The female can store the sperm for

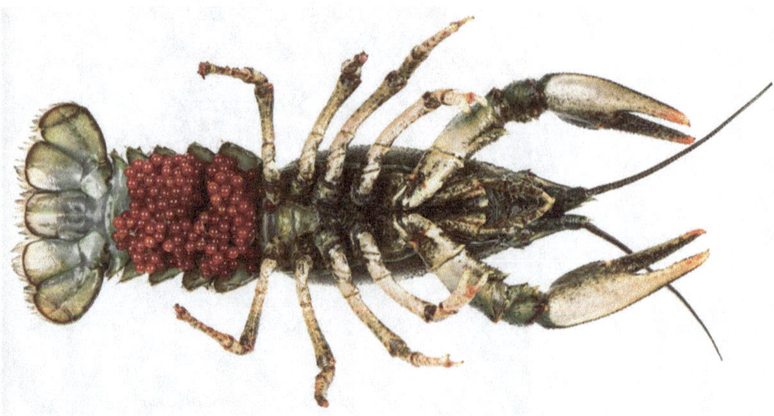

Fig. 6.2 Crayfish with roe ("berries")

fertilization for up to 6 months after mating. The number of eggs varies but for a large crayfish can be up to 500 eggs. When the fertilized eggs mature, they become more brownish. If they are unfertilized, they become more orange. Fertilized eggs hatch after 3–4 weeks. There is no information available on the ingredients of crayfish roe. Crayfish roe can be eaten raw and salted like other types of kaviar but can also add flavor and texture to soups and sauces. The flavor is very mild.

6.3 Lobster

There are two types of lobster, the European lobster (*Homarus gammarus*), which inhabits the west coast of Europe and in the Mediterranean, and the American or Canadian lobster (*Homarus americanus*), which lives on the Atlantic coast of North America. The European lobster is fished from the North Sea to the Baltic Sea and in certain fjords. The European lobster is also called the black lobster because, unlike the reddish-brown American lobster, it is dark and blackish blue.

The immature roe of a lobster are seen as a dark green jellylike mass, called the coral, under the carapace, and as mentioned earlier, the dark color is determined by a protein complex called ovoverdin. However, the color can depend on the lobster's food. After mating, the female lobster stores the male's sperm until the eggs are fertilized. The eggs lie inside the lobster for 9–12 months before they are released and stuck between the small swimlets on the underside of the tail. The eggs are 1–2 mm. They sit between the swimlets for another 9–12 months before hatching. The small, hatched larvae float in the upper water layer of the ocean together with other plankton for 4–6 weeks, after which the survivors sink to the bottom and can take up a life on the sea floor. A female can produce 5000–160,000 eggs every 2 years, depending on the size of the animal.

Lobster roe have about 20% protein, 4% fat, and quite small amounts of cholesterol (0.1%). The ω-3/ω-6 ratio is about 5.

Lobster roe are not a commercial product, but if you get hold of a lobster with "berries," i.e., with lumps of black eggs under the tail, they can be eaten as they are, possibly lightly salted like caviar. The immature roe, the coral, can be used in a sauce, a bisque, or mixed in butter, and this applies to both the raw black and the cooked red roe. When cooked or grilled, the immature roe becomes firm, red, and slightly waxy.

6.4 Norway Lobster

Norway lobster (*Nephrops norvegicus*) are a deep-water species. It is one of the most important shellfish resources in the northeast Atlantic and is the most commercially valuable lobster product in the world. Norway lobsters are found from Iceland across the Mediterranean coast to Morocco. As with lobsters, the immature

Fig. 6.3 Boiled Norway lobster with immature roe (coral) under the carapace

roe or coral of the Norway lobster are found as a dark mass under the carapace which turns red upon heating (cf. Fig. 6.3). The developed eggs are secreted under the abdomen and migrate out under the tail and are placed in glued lumps ("berries") between the small swimlets. The eggs are very dark and typically black but can also be lighter and greenish, and they have a diameter of about 1.5 mm, and their shape is like a flat ellipsoid.

A female Norway lobster can carry around 1000 eggs, depending on its size. Analysis of the immature eggs has shown that they contain approximately 63% water, 25% protein, 10% fat, and 1–2% carbohydrate. There is a large predominance of polyunsaturated fatty acids, including super-unsaturated fatty acids. The ω-3/ω-6 ratio is around 6. However, this decreases slightly as the eggs mature.

Extensive studies have been carried out on Norway lobster roe at different stages of maturation in order to gain an understanding of what determines the development of the larvae. It turns out that although the main component of the roe is protein, it is mainly fats that are used as an energy source during the formation of the embryo in the egg, and the egg therefore loses fat during maturation. On the other hand, the amount of free amino acids increases considerably, especially the umami-tasting glutamate, which increases by a factor of four. At the same time, the egg grows, and the water content increases by 20%. This means that the umami taste of the roe effectively increases as the eggs mature.

6.5 Shrimp and Prawn

Shrimp and prawn constitute two large families, the true shrimps (Caridea), which include, e.g., Baltic shrimp, caridean shrimp, and deep-water shrimp, and the penaeid shrimps (prawns; Dendrobranchiata), which have a different gill structure, such as tiger prawns. Deep-water shrimps (*Pandalus borealis)* from the cold North Atlantic seas make up by far the largest part of the world's shrimp catch.

Similar to other crustaceans, there is very little information about the ingredients in shrimp and prawn roe. The water content is around 70%, the fat content is 4–7%, and the ω-3/ω-6 ratio varies from 1 to 7. A number of Arctic deep-water shrimp have been found to have a water content of 70% and a fairly small fat content of around 5%, of which more than a third are super-unsaturated fatty acids. The cholesterol content varies quite a bit, from 0.8% in Baltic shrimp to just 0.11% in some tropical shrimp. Like other types of roe, shrimp roe (*ebiko*) are a sought-after delicacy in Japan.

Since the head is usually removed from shrimp when they are peeled, the immature coral under the shell is rarely seen. When heated, the coral turns red (cf. Fig. 6.4). Collections of eggs are frequently seen between the small swimlets under the tail body (cf. Fig. 6.5) on whole shrimp when sold live. It can also happen that there is roe on deep-frozen cooked shrimp that have not been peeled.

Roe on shrimp can have different colors depending on the species. The main components of the pigments are the red and orange carotenoids, astaxanthin, and doradexanthin, which only become clearly visible when the shrimp are cooked. The fresh roe can be grayish, yellow, gray blue, or very dark and almost black. The individual eggs are very small, approximately 1.0–1.5 mm in diameter, depending on the species, and thus smaller than roe from flying fish. Eggs from rockpool shrimp (*Palaemon elegans*) are only 0.8 mm in size. The number of eggs varies for different

Fig. 6.4 Grilled giant prawn, where the roe (coral) has turned red

Fig. 6.5 Small shrimp with eggs between its swimlets

Fig. 6.6 Roe from boiled Baltic shrimp with eggs of different colors

species from a few hundred to 1000 eggs. Another delicate shrimp is the small Baltic shrimp (*Palaemon adspersus)* that lives in inner fjords in Scandinavia and has a delicious roe (cf. Fig. 6.6).

When Peeling Baltic Shrimp

It is a lot of work to peel your own small Baltic shrimp, and it takes a lot for a single shrimp meal. Most people have probably noticed that home-peeled shrimp tastes best. It is certain that the pleasure of peeling your own shrimp influences the taste assessment. However, there may be a more objective reason for this, namely, if the home-peeled shrimp carry roe, and you avoid rinsing the roe from the peeled tails. The shrimp roe give the shrimp extra umami taste, so do not rinse the peeled shrimp!

Chapter 7
Roe from Mollusks

The phylum Mollusca arose during the so-called Cambrian explosion about 542 million years ago, during which time, over a period of 50 million years, new life forms emerged, including all the major animal orders that we have today. There are at least 85,000 living species of mollusks and at least as many extinct ones. Some mollusks have outer shells like clams and snails, others have rudimentary internal skeletons like ten-armed squids, and others, like eight-armed octopus, have no other hard parts than their beaks. All of these mollusks reproduce with eggs and therefore have roe.

7.1 Cephalopods (Cuttlefish, Octopus, and Squid)

There are about 800 different species of extant cephalopods in the world's oceans. It is the fleshy parts of cephalopods, such as the mantle, arms, and tentacles, that are used in cooking, and the intestines and reproductive organs are usually thrown away. Cephalopod roe are therefore not usually a commercial item in the West, and cephalopod eggs are used to a much lesser extent in gastronomy than roe from fish, crustaceans, mussels, and echinoderms. Cephalopod roe appear in Thai cuisine, and the gonads of certain cuttlefish are eaten in Southern Europe.

In contrast to the rather small eggs of most fish, crustaceans, and echinoderms, the eggs of some cephalopods are much larger, often ten times larger in diameter. They also contain much more water. The eggs are soft and slightly flabby in consistency.

A specialty of Andalusia is *huevos de choco*, which are the nidamental glands of large cuttlefish of the species *Sepia officinalis* (cf. Fig. 7.1). The coral is seen at the tip of these glands. The nidamental glands, which produce gelling agents to hold together and harden the eggs of the females, are a gastronomic and not cheap specialty. They require almost no preparation. They are prepared either grilled or in a marinade with olive oil, garlic, and parsley.

O. G. Mouritsen, K. Styrbæk, *Roe and Roe Gastronomy*,
https://doi.org/10.1007/978-3-032-13142-3_7

Fig. 7.1 The nidamental glands, *huevos de choco*, from large cuttlefish of the species *Sepia officinalis*

Salted roe from octopus are considered a delicacy in Greece. It is not a commercial product but is made as a homemade manufacture. The same applies to dried roe (cf. Sect 11.4) *bottarga* from octopus roe (cf. Fig. 7.2), which have a milder and less bitter taste than *bottarga* made from fish roe.

On a Field Trip: Octopus Bottarga at a Portuguese Restaurant

Chef and gastronome André Magalhães is a legend in gastronomy not only in Portugal, where he runs his celebrated little restaurant, Taberna Rua de las Flores in Lisbon, but also in Africa and Asia, where he has worked to spread knowledge about local gastronomy, for example, through film projects about women describing regional recipes, which are brought to mind through music and song. At André's restaurant, you cannot book a table, and the tables, chairs, plates, and cutlery are all recycled. Taste is paramount, and the quality and inventiveness are so high that Michelin chefs flock to eat at André's table.

One of the authors (Ole) has been lucky enough to eat at Taberna Rua de las Flores a few times, and although he has encountered many unusual specialties there, it was a surprise to be presented with octopus roe *bottarga* made according to a traditional recipe, which involves salting and air-drying the fresh roe, which is then pressed to a suitable consistency. The resulting *bottarga* is then cleaned, rubbed with olive oil, and now has a long shelf life. André's octopus *bottarga* is shaped like a puck, very firm, and grayish on the surface (cf. Fig. 7.2). This type of *bottarga* is eaten by first heating it over charcoal or with a gas burner, so that it becomes softer and can be cut into thin slices over, for example, a green salad with a little olive oil, possibly toasted almonds, or finely chopped garlic.

Fig. 7.2 *Bottarga* made from octopus roe. Prepared by André Magalhães

7.1.1 Octopuses

Octopuses are octopods and have eight arms. They have been studied more thoroughly than the ten-armed decapods, squid, and cuttlefish, and data is available for a wide range of different octopus species. The eggs of *Octopus vulgaris* are whitish

Fig. 7.3 Octopus eggs on the left and cuttlefish eggs on the right

and somewhat smaller than those of cuttlefish and are the size of a grain of rice (cf. Fig. 7.3). They are shaped like ellipsoids, about 2.5 mm on the long side and about 1 mm on the short side. The water content of the eggs is about 75%, the protein content about 15%, and the fat content 2–4%, depending on the species. The ω-3/ω-6 ratio is quite high and around 5, and the cholesterol content is similar to that of fish eggs. A single egg weighs only about 1.4 mg.

7.1.2 Squid and Cuttlefish

Decapods are divided into two classes: torpedo-shaped squid (e.g., *Loligo* spp.) and cuttlefish (e.g., *Sepia* spp.). The water content of the roe of cuttlefish (about 90%) is somewhat higher than that of squid, which therefore also has a significantly higher protein and fat content. The ω-3/ω-6 ratio is high, especially for squid, where it is 28. Squid eggs are white or yellow and small (cf. Fig. 7.4), about 2 mm and slightly elongated. Cuttlefish eggs are much larger, about 13 mm, and a single egg weighs 0.75 g and is oval in shape (cf. Fig. 7.3). There does not seem to be any widespread gastronomic tradition of eating roe from squid and cuttlefish.

Fig. 7.4 Fresh roe in the mantle of squid cut across

7.2 Mussels

Mussels belong to a class (Bivalvia) of mollusks with about 20,000 different species, most of which live in saltwater. Examples include blue mussels, cockles, razor clams, scallops, and oysters. Mussels are filter feeders, and their food consists of plankton, microalgae, and nutrients as well as small particles of organic matter dissolved in the water. Their taste, and this applies to muscular tissue and especially to the intestines and gonads (coral), bears the imprint of the mussels' environment ("merroir"). Because mussels are at the bottom of the food web, they are a relatively sustainable food, and it has recently been shown that mussels in aquaculture and cultivated seaweed are the two most sustainable marine foods.

Reproduction in mussels occurs when the males release sperm into the water, and the females take in the sperm together with the seawater when they filter for food. The eggs remain in the female until they are fertilized, and the larvae develop inside the female, after which they are released into the water. They then attach themselves as so-called spat to the seabed or other solid objects.

When eating mussels, one does not usually distinguish between the different parts of a mussel's interior, other than possibly between the fleshy adductor and the rest. For large mussel species, the foot and siphon are also in focus, but for, e.g., blue mussels (*Mytilus edulis*) and oysters (*Ostrea edulis* and *Crassostrea gigas*), the entire inside of the shells is eaten as a whole, and any roe are not given any special attention, unless the mussels are fat and close to spawning.

The individual eggs in the roe of mussels are very small, so one will not normally be able to distinguish them but perceive the gonads as a soft, homogeneous mass. The fat content is quite low, around a few percent, and the ω-3/ω-6 ratio varies from 2 to 10, depending on the species.

For scallops (members of the family Pectinidae), something special applies, because this mussel is free-swimming and therefore has a large swimming

Fig. 7.5 Scallop with orange roe (the coral) seen next to the adductor muscle

(adductor) muscle (cf. Fig. 7.5). It is this large, white muscle that is considered as food, and the rest of the viscera are not eaten. However, if the scallop has roe, the roe sac is seen as a grayish-white or orange-colored appendage to the adductor muscle (cf. Fig. 7.5). This is the coral and it has a soft and very delicate texture.

7.3 Snails Egg: White Kaviar

Land snails are mollusks that are widespread in the Earth's temperate and subtropical zones. Their meat is part of a number of food cultures, for example, in France and Poland, where the meat of the escargot (*Helix pomatia*) is a gastronomic specialty. Escargot is the largest European land snail, and its eggs, which resemble small white pearls of about 3–6 mm (cf. Fig. 7.6), have been called "white kaviar." They are considered a special delicacy, which has also been called "escargot pearls" or Aphrodite's pearls. The latter name refers to the fact that the snail eggs have been attributed with properties that promote love.

There are also some slightly smaller, gray snail species, *Cornu aspersum* ssp. (formerly called *Helix aspersa* ssp.), which are a type of common garden snail. These are mainly *Cornu aspersum maxima* and *Cornu aspersum aspersum*, and both species are cultivated, e.g., in Poland and Chile, for both meat and egg production. Snail farming goes back a long way in history, and snail farms existed during the Roman Empire and are mentioned by the historian Pliny the Elder (23–79 AD)

Fig. 7.6 Eggs from land snail

around 50 AD. There are also Roman descriptions of how snails could be fattened on emmer flour and grape syrup.

A land snail becomes mature to lay eggs after 4–7 months, and it typically lays 100 eggs, which correspond to 4–5 grams. This means that it takes 25,000 eggs to produce 1 kilo of "white kaviar." The eggs are loose and are laid one at a time by the snail. Since they have to be carefully picked up by hand and sorted, it is clear that "white kaviar" must be an expensive product.

Snail eggs are perfectly spherical and have a dull and milky white color. They have a firmer and silkier texture than fish roe, e.g., from salmon and sturgeon, but are comparable in size. When you chew on snail eggs, they explode like certain fish eggs but with a more pronounced explosion. The taste and aroma are slightly earthy with notes of moss, mushrooms, nuts, oak, and wild asparagus.

Apart from minor variations, snail eggs contain 87% water, 4–6% protein, and apparently only a little fat, although there are large uncertainties in the literature. It is significant that the water content is higher than in fish roe and crustacean roe. In contrast to eggs from marine animals, the fat in the eggs of land snails consists predominantly (about 75%) of saturated fatty acids, and there is very little of the polyunsaturated fatty acids. Thus, the protein and fat content is significantly lower than in fish roe. However, snail roe, in contrast to fish roe, have a high carbohydrate content, about 4–5%, and there are good amounts of the minerals calcium, copper, and iron.

The energy content of snail eggs is very low, about 185 kJ/100 g, which is ten times less than in fish roe. The difference is due to the small fat content in snail eggs, and that energy for the new individuals comes from carbohydrates instead. All in all, this means that snail eggs are not a good source of nutrition.

Snail kaviar products are pasteurized or preserved in a salt solution. The eggs are stored cold at about 10 °C, and some prefer to eat them at room temperature. Snail kaviar is used as a decoration on various dishes, just like fish eggs, and this is where the aesthetic effect of the white pearls is particularly evident.

The decorative value of snail egg was used by the Danish star chef, Kenneth Toft-Hansen, in a decorative symphony that won him the Bocuse d'Or Gold Medal in 2019 with a so-called decorative symphony. This symphony consisted of "white kaviar," apple, and smoked cream cheese (cf. Fig. 7.7). Kenneth chose the white snail eggs because these eggs are good at absorbing flavors from marinades of, for example, tarragon and garlic oil. By also smoking the eggs, he gave them both smoky flavor notes and a slightly yellowish visual appearance like real pearls.

Fig. 7.7 A section of Kenneth Toft-Hansen's Bocuse d'Or decorative symphony 2019 with "white kaviar" of snail eggs, apple, and smoked cream cheese

Chapter 8
Roe from Echinoderms

Echinoderms are a phylum (Echinodermata) of invertebrates, of which there are about 6000 different living species, which include, for example, the classes of sea urchins, starfish, and sea cucumbers. They are characterized by having an external armor in the form of a calcareous shell or a leathery thick skin. You have to get behind the thick armor to find the delicacies in the form of roe.

8.1 Sea Cucumber

Sea cucumbers are a class (Holothuroidea) of echinoderms with many families, which are used as food in various parts of the world, especially in East and Southeast Asia. The world market is 48,000 tons (2018). It is mainly species such as *Stichopus* spp., where viscera and gonads are used.

There is little available information about the ingredients in roe from sea cucumbers. The fat content in sea cucumber gonads is 5–10%. There is a predominance of polyunsaturated fatty acids, but the content of polyunsaturated fatty acids decreases significantly in fermented products.

Viscera from sea cucumbers find their way into salted and fermented delicacies (*konowata*). During the spawning season, the sea cucumber's gonads swell in a place near the sea cucumber's mouth, which is why the roe are called *konoko* or *kuchiko* in Japan, meaning "mouth children." In Japan, a delicacy is made from sea cucumber gonads that are dried into thin wafers (cf. Fig. 8.1), each of which requires 10–15 kg of sea cucumbers. *Kuchiko* is therefore a rather expensive delicacy. *Kuchiko* is eaten as is, salted, marinated, or roasted, and is used in soups.

O. G. Mouritsen, K. Styrbæk, *Roe and Roe Gastronomy*,
https://doi.org/10.1007/978-3-032-13142-3_8

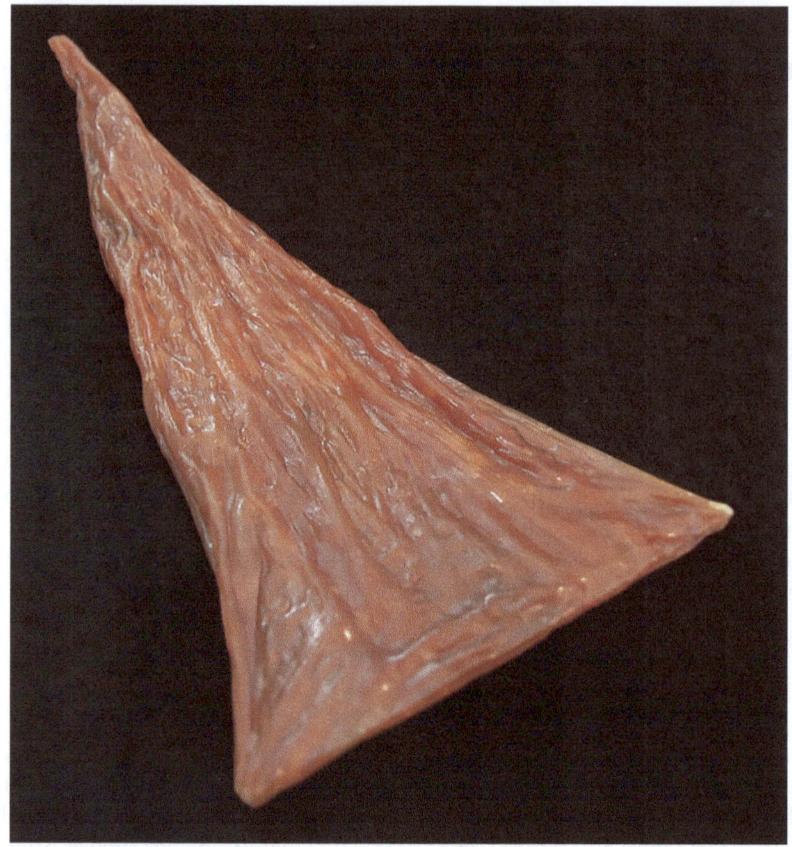

Fig. 8.1 Dried sea cucumber gonads, *kuchiko*. Forty sea cucumbers have been used to produce this piece being 7 cm wide

8.2 Sea Urchin

Sea urchins are a group of echinoderms that include 950 different species, and although all species are believed to be edible, only a handful are exploited commercially, such as *Strongylocentrotus* spp. and the red Chilean sea urchin *Loxechinus albus*. Together, these two species make up the majority of the market. Other species include *Strongylocentrotus nudus* and *Strongylocentrotus intermedius* in Japan and *Paracentrotus lividus* in Italy and southern France. Still another species, *Psammechinus miliaris*, also known as the green sea urchin, is found along the Atlantic coast around Brittany and the English Channel. This species is particularly sought after. In New Zealand, where the Māori have a rich tradition of eating sea urchin roe, it is the species *Evechinus chloroticus* that has been caught.

Sea urchins are harvested both in the wild and can be grown in aquaculture, which, however, constitutes a vanishing part of the market. They are found in most of the world's oceans but are difficult to harvest, and this is done with the help of scuba divers or traditionally, e.g., in South Korea, by free diving down to a depth of 15 meters. In some places in shallow water, visible sea urchins are also scraped up by hand from the seabed with a special rake. Since these classic harvesting methods are expensive and laborious, experiments have been carried out in Norway with using a motorized rover that can be remotely controlled to work on the seabed to collect sea urchins and shellfish such as scallops.

Chile has traditionally been the world's main producer of sea urchins but now with declining production. Other producing countries are Japan, the USA (California and Maine), and Canada (British Columbia). On the North American west coast, the large red sea urchin *Strongylocentrotus franciscanus* is harvested in particular. The annual global production of sea urchin roe, which has been decreasing and fluctuating over the past decades, is now around 75,000 tons (2018), the vast majority of which is consumed in Japan, followed by France and South Korea. In Europe, the largest consumers are France, Italy, and Spain.

The only part of the sea urchin which is eaten is the so-called roe (cf. Fig. 8.2), which are actually the combined, yellowish-brown reproductive organs (gonads), i.e., testicles and ovaries, which can be difficult to distinguish from each other, but which take up two thirds of the intestines. In wild sea urchins, roe can make up to 25% of the weight. In aquaculture, it is possible to increase the weight of the roe by 40%, but the color and flavor do not seem to be affected and are mostly determined by the sea urchins' basic diet of algae, seaweed, and possibly small animals.

Fig. 8.2 Sea urchin roe in the shell

Sea Urchins, Starfish, Sea Otters, Seaweed, and Marine Ecosystems

The balance between different species in marine ecosystems is delicate and complex. As an example involving four species, in 2014, a rapid decline in the large kelp forests of macrokelp (*Macrocystis pyrifera*) was observed off the coast of California, leaving large barren patches on the seabed but still a mosaic of healthy, smaller patches of seaweed. At the same time, an increase in the number of purple sea urchins (*Strongylocentrotus purpuratus*) and sea otters (*Enhydra lutris nereis*) was observed. In contrast, populations of starfish (*Pycnopodia helianthoides*) were declining. Sea urchins eat seaweed, and starfish and sea otters eat sea urchins. The delicate balance between the four dependent species had been disrupted, and large areas of kelp forests were completely grazed by the sea urchins. One would think that the growing population of sea otters could restore the balance, but this only happened in those spots where there was still some kelp forest left. The reason for this was that the sea otters were not interested in eating the many starving sea urchins in the barren areas, because they contained only a little roe.

Now initiatives are being launched to regulate the population of sea urchins, strengthen the growth of the kelp forests, and ensure better roe in the sea urchins. The international company Urchinomics is a leader in this area with the aim of building local markets for the consumption of sea urchin roe.

The gonads are the sea urchin's most important organ for storing nutrients. After spawning, very little is left; the gonads have shrunk and are barely visible, and the taste has become very bitter.

Sea urchin roe contain about 16% protein and 8% fat. The cholesterol content is high, around 0.3–0.5%. The high fat content of the roe and the fact that the eggs are very small, around 0.8 mm, make the roe creamy. This creaminess is also behind the intense sea flavor of salt, iodine, and bromine in the raw roe, because the fats provide a coating mouthfeel that maintains the intensity of the flavor for a long time. The consistency of the roe is almost like an egg cream. The roe can therefore also be used to thicken a sauce or a soup. The strong marine flavor of the roe is mostly due to the fact that sea urchins eat seaweed, mainly brown seaweed species.

The special taste of sea urchin roe is due to the very high content of umami-tasting substances, both free glutamate and several different free nucleotides, which provide umami synergy. This means that sea urchin roe are the type of roe that has the most intense umami taste.

Sea urchin roe are considered a great delicacy, not least in Japan, where it is called *uni*, but there is also a tradition of eating the roe in France, Portugal, and China. The roe are usually sold fresh, and it should preferably not be frozen and therefore transported fresh. However, you can also get the roe pasteurized in glass or cans, but as such it is far less delicate than the fresh roe, and the texture is more liquid. The roe are also used for some fermented products (*neri uni*), and in some versions it is salted, dried, or washed with alcohol (*doro uni*).

The quality of the roe is often judged by its color, and the bright and intensely orange-colored roe are most appreciated. The light roe also has the mildest and most delicate taste, in contrast to the roe, which is dark brown and has an almost too strong marine aroma and taste of iodine in particular. Because sea urchin roe have large amounts of free glutamate and free nucleotides (IMP), a very strong umami synergy occurs in the taste. This strong taste is used, among other things, in Chilean cuisine to add flavor to ceviche dishes.

The quality of the roe depends on the species and time of harvest. One of the species that is attributed to the finest quality of roe is the so-called red sea urchin (*Strongylocentrotus intermedius*). In Japan, the red sea urchin is harvested around the northern island of Hokkaidō, and the roe are yellow orange in color and has a mild, sweet, and nutty taste as well as a long aftertaste. In Japan, this roe is called Ezo Bafun, and its particularly rich umami taste is attributed to the sea urchin foraging on the large Japanese seaweed *konbu (Saccharina japonica),* which contains very large amounts of free glutamate, which gives the umami taste. *Strongylocentrotus intermedius* is also harvested in California. The purple sea urchin, *Strongylocentrotus purpuratus*, is fished in Japan around Honshu but also on the northeastern coast of the USA in Maine. Its roe are also sweet but have a browner color than that of the other two species.

In many countries, sea urchin roe are eaten raw with a spoon directly from the shell, where the bottom is cut off, exposing the fine, five-pointed symmetrical arrangement of the gonads (cf. Fig. 8.2). Sea urchin roe are a valued delicacy in a number of countries besides Japan, such as Chile, South Korea, and China. The roe are eaten raw in Lebanon, and in the Philippines, the whole sea urchin in its shell is cooked either by steaming, roasting, or heating it over an open fire, and the contents are eaten with a little juice from the citrus fruit calamansi. Chile is the world's largest producer of sea urchin roe, and here the roe are traditionally eaten raw marinated with lemon, onion, and olive oil.

In Europe, it is mainly in the Mediterranean countries where there is a tradition of fishing and eating sea urchins, and this applies especially to Italy, Spain, and France. In France, sea urchin roe are found in a wide range of dishes such as soufflés, scrambled eggs, omelettes, fish soups, and sauces such as mayonnaise and hollandaise. In soups and sauces, the roe also act as a thickener and to give a yellow-brown color. There are also French varieties of *taramasalata*, where sea urchin roe can appear together with different types of fish roe. A traditional Provençal preparation is called *oursinado*, which is made by crushing the roe in a mortar and adding a little olive oil along the way, forming an emulsion like a kind of mayonnaise. *Oursinado* is eaten with boiled potatoes or bread. In Italy, the roe are used with pasta dishes and even on pizza. In Spain, sea urchin roe can be found in stews and soups or as a side dish for egg dishes.

While sea urchin fishing in Southern Europe is about the warm-water species *Paracentrotus lividus*, in Northern Europe it is *Strongylocentrotus droebachiensis* that is found. However, in the Nordic countries, there is no tradition of catching and eating sea urchins, but there are good occurrences of the species *Strongylocentrotus droebachiensis*, and there is a small, developing fishery of this species in Norway, the Faroe Islands, and Iceland.

Sea urchin roe are a very fragile and delicate food that easily loses freshness and firmness after it is removed from the shell. By far the best eating quality is therefore achieved by eating the fresh roe out of the shell. When transporting roe that have been removed from the shell, alum salt is sometimes used, which preserves the freshness and shape of the roe but can give a metallic taste. Newer preservation methods use degassed water in which some nitrogen is dissolved, which helps to prevent oxidation of the fats.

Sea Urchin Roe in Antiquity

The ancient Greeks are said to have enjoyed raw sea urchin roe, which were mixed with wine vinegar, sweet wine, honey, parsley, and mint. The Romans, on the other hand, cooked the sea urchin over the hearth in the shell, mixed with oil, *garum* fish sauce, sweet wine, and black pepper. Other Roman recipes prescribe balsam leaves, bay leaves, and various herbs for sea urchin roe.

***Uni:* Sea Urchin Roe in Japan**

Most of the world's production of sea urchin roe, *uni*, ends up on the dining tables of Japan, which accounts for around 90% of world consumption. This large consumption is rooted in a long and strong culinary tradition throughout Japan, which dates back to prehistoric times and can be documented with certainty to 718 BC. Back in the 1980s, Japan was considered the world's largest producer of sea urchins, but the harvest has since declined, and Japanese demand can only be met by massive imports. There are over 100 different species of sea urchins in Japan. Six of them are harvested commercially, and *Strongylocentrotus nudus* makes up 40–45% of the market.

The taste of Japanese sea urchins depends on the season, where they are harvested, and what they have eaten. The quality is judged primarily by color, the density of the individual eggs in the roe, and the condition of the blood vessels, which signal freshness. To illustrate the incredibly refined Japanese food culture around sea urchin roe, where each region's special qualities of the local sea urchins are in focus, follows here a description of the different species and their culinary advantages.

The most sought-after sea urchins come from the cold waters around the northern island of Hokkaidō, where the species Ezo Bafun *uni* (*Strongylocentrotus intermedius*) and Kita Murasaki *uni* (*Heliocidaris crassispina*) can be harvested twice a year. In these nutrient-rich waters grows the brown seaweed *konbu* (*Saccharina japonica*), known for its umami taste, which is the sea urchin's preferred food. Kita Murasaki *uni* is particularly sought after because its gonads are large and very light and have a creamy, sweet flavor. The roe from Ezo Bafun *uni* is considered the creamiest.

Further south, in the waters around the main island of Honshu, the two species Bafun *uni* (*Hemicentrotus pulcherrimus*) and Murasaki *uni*

(*Heliocidaris crassispina*) are harvested. Both species feed primarily on seaweed, and in this area, it is not *konbu* but *wakame* (*Undaria pinnatifida*) that gives the roe a less delicate flavor. Murasaki *uni* is considered an average-to good-quality product with a slightly fishy aftertaste. Bafun *uni* produces roe that are darker than Murasaki *uni*, and its flavor is stronger and slightly more bitter, which is because the sea urchins forage in deeper water.

On the southern island Kyushu, it is mainly the red sea urchin, *aka uni* (*Pseudocentrotus depressus*), that is harvested for roe. The water here is warmer than in the north and less nutritious, and the sea urchins do not grow as large. The food is still *wakame*, but also crustaceans and small fish, which give the roe a rather bitter and fishy taste. This roe are also less creamy and are considered to be of lower quality than the roe types from more northern waters.

The Japanese especially prefer sea urchin roe that are yellow orange and have a creamy texture like an egg custard. The roe have optimal gastronomic quality just before the spawning season, when the gonads are fully developed. The best season for Murasaki *uni* is July to August, September to October for Kita Murasaki *uni*, October to November for Bafun *uni*, and December to February for Ezo Bafun *uni*. Outside these seasons, the Japanese import *uni* from abroad.

Traditional culinary uses of *uni* in Japan include raw as *sashimi*, as a topping on sushi (cf. Fig. 8.3), or placed on top of a bowl of hot rice, *uni donburi*, which is a typical breakfast dish in Hokkaidō. *Chawan mushi* is a traditional egg custard made with *dashi* stock in which the roe are placed. *Uni* is also used in hot dishes, such as hotpots like *ichigo-ni*, which are eaten at New Year's celebrations.

Fig. 8.3 *Nigiri-sushi* with sea urchin roe (*uni*) on top

Sea Urchin Festival in Portugal

Every year, an international sea urchin festival, Festival Internacional do Ouriço-do-mar, is held in the small Portuguese coastal town of Ericeira. It is said that the name of the town is derived from the expression *ouriceira*, which means "sea urchin land." At the festival, chefs from near and far demonstrate how to prepare sea urchin roe. At this festival, a survey was conducted in 2019 to find out people's attitudes and interest in eating sea urchin. Although Portugal is the world's third largest consumer of seafood, the Portuguese do not have a great tradition of eating sea urchin. The results of the survey showed that while many people are prepared to eat sea urchin in a restaurant, the desire to buy the product in the supermarket is significantly lower, and knowledge of eating sea urchin roe is mainly from restaurant visits. In the Alentejo province in southeastern Portugal, there is a traditional dish where sea urchins are cooked in their own shells by placing burning pinecones around the shell.

8.3 Starfish

Starfish are star-shaped marine animals that belong to the class Asteroidea. Due to their hard surface with lots of small spikes and an often-bitter taste, starfish are only eaten in a few places in the world, such as China, Japan, and Micronesia.

Some species of starfish contain toxic substances, and the thick spiny skin with lots of small lime spikes does not invite a meal. But just like sea urchins, there is a soft interior in the starfish in the form of the reproductive organs, the gonads, which lie in channels out into the five arms (cf. Fig. 8.4). Toward the time of spawning, the content of umami-tasting free glutamate in the gonads grows to a fairly high level.

The gonads of starfish can be scraped out with a small spoon after making an incision along the top of each arm. There is not much in each animal. The roe have a creamy and fatty mouthfeel and can be lightly fried, for example. The protein content is around 12%, and the fat content is relatively high, around 10%.

Fig. 8.4 Starfish with gonads exposed by an incision in each arm

Chapter 9
Artificial Fish Eggs and Kaviar Substitutes

Kaviar substitutes can be made from roe from fish species other than sturgeon and even contain or be produced as granulated products of fish or crab meat, microalgae, seaweed, or vegetables. These products are often dyed black to imitate real caviar. Eggs from snails are also marketed as snail caviar or snail kaviar.

None of these products may be marketed as caviar but as kaviar or caviar substitutes. However, the marketing is often done in a way that signals a certain degree of luxury product.

9.1 Kaviar from Fish Meat

The Spanish products Avruga and Moluga (cf. Fig. 9.1) are little pearls and should not really be called kaviar because they are processed products based on whole herring (*Clupea harengus*) and possibly herring roe if the herring used contains roe. However, they are marketed to some extent as a kaviar substitute, but they are in a way a product that has a fine gastronomic value in itself. At the same time, they are an example of a good and relatively sustainable use of herring. The products are dyed black with *Sepia* ink, and they are smoked. The difference between Avruga and Moluga is that Moluga is a luxury product made with North Sea herring with a higher fat content than Avruga, which makes the pearls creamier.

Avruga and Moluga were developed by the small Spanish company Pescaviar in Madrid with the aim of producing a product that has some of the textural characteristics of caviar, including creamy, buttery, and rounded in taste. The taste is mild and less fishy than real caviar. The individual beads are around 2 mm in diameter. Avruga and Moluga do not taste like caviar or like untreated herring roe or herring fillet, which can have a fishy and often bitter taste.

O. G. Mouritsen, K. Styrbæk, *Roe and Roe Gastronomy*,
https://doi.org/10.1007/978-3-032-13142-3_9

Fig. 9.1 Selection of caviar made from fish meat: Avruga, Moluga, and Anchoviar

There is a similar product, Anchovia (cf. Fig. 9.1), which is made from anchovy (*Engraulis encrasicolus)*, which is not smoked. The anchovy beads are also 2 mm in size. The products from Pescaviar are also available in versions that are MSC (Marine Stewardship Council) certified.

9.2 Seaweed Kaviar

Various marine seaweed species have a taste and texture that can be compared to roe in some ways. They can therefore be used in the kitchen in the same ways as fish roe, especially as a garnish. In addition, seaweed contain some special and very complex carbohydrates that are good at binding water, and both powdered seaweed and seaweed extracts are therefore used as gelling agents (so-called hydrogels) and thickeners. Brown seaweed species contain the substance alginate, which can form strong gels in the presence of calcium ions. Under special circumstances, such gels can be made in the form of a spherical shell, where a liquid can be enclosed inside, for example, with a special taste, color, or aroma. The spherical shells can have different sizes, for example, like fish roe.

9.2.1 Green Kaviar

Sea grapes (*Caulerpa lentillifera*) are a green seaweed species that resemble clusters of small grapes (cf. Fig. 9.2). Sea grapes are eaten raw as a refreshing sea salad. In the Philippines, sea grapes are called "green caviar," and in Japan they are called *umibudō*, which means sea grapes. Sea grapes have a crispy texture and a juicy and refreshing mouthfeel. The special sensation that occurs when you crush sea grapes between your teeth has even been given its own onomatopoeic expression in Japanese, *puchi-puchi*. In the Japanese archipelago of Okinawa, where *umibudō* is traditionally eaten, this green caviar is attributed with the ability to give a long life.

Fig. 9.2 Sea grapes from the green seaweed species *Caulerpa lentillifera*

9.2.2 Cavi-Art

Cavi-art is a Danish trademark that plays on the similarity of names with caviar. It was originally a product that was intended to be a replacement for fish kaviar, but in recent years it has acquired a value in itself as an accessory of small beads or pearls that can have different flavors and colors, so that they can also be included in desserts, for example. The mouthfeel of Cavi-art is like that of fish eggs that burst and the liquid leaks out when you crush the pearls between your teeth. The interesting thing is that the pearls can be tasteless while they are still whole, but as soon as they break, there is an explosion of flavor and aroma. The big difference, however, is that the Cavi-art pearls do not have a creamy and melting mouthfeel like fish roe, because they do not contain fats and oils.

The story behind Cavi-art began in 1988, when Danish inventor Jens Møller discovered that certain enzymes added to seaweed caused the formation of some small spheres. Jens Møller was quick to see the similarity with fish roe, and after a few years of development, he was able to take out a patent in 1994 and, based on his invention, in 1995 started a company to manufacture a very unusual product, which he named Cavi-art.

Cavi-art are small spheres made of seaweed, and as the name suggests, these spheres resemble fish eggs (cf. Fig. 9.3). And the most important thing is the mouthfeel. The small Cavi-art spheres consist of a firm and elastic membrane of gelled alginate from the seaweed, and inside there is a liquid mass that can be given different colors and flavors depending on what you want to use the spheres for. It can be as a replacement for kaviar, where the spheres have a marine aroma, but it can also be used in sweet cuisine, for example, Cavi-art pearls with encapsulated fruit juice. The pearls can be used for both cold and hot dishes, and the product is not damaged by heating. The colors do not bleed, which gives Cavi-art advantages as a garnish for boiled eggs, for example.

Fig. 9.3 Cavi-art, a kaviar substitute made by an extract from brown seaweed species

Cavi-art can be seen as a clever application of a process called spherification as described below. Interestingly, Jens Møller made his invention and applied it to food before avant-garde chefs believe they "invented" the use of alginate to make spheres in the modernist and molecular cuisine.

Spherification: Make Your Own "Fish Eggs" and Kaviar

It is quite easy to make your own "fish eggs" and kaviar from spheres made by spherification as described below. Making small spheres is straightforward, but the slightly larger ones require some practice. In addition, you can also easily enclose liquids with different flavors and colors inside the eggs (cf. Fig. 9.4), as long as you do not use juices that are too acidic.

By using alginate (sodium alginate, E-401), which is extracted from brown seaweed, it is possible to form small spherical shells (pearls) with a liquid interior. This results in a fun texture that provides surprising challenges to the mouthfeel. It can be a crunch, like small fish roe, or a yielding elastic behavior first and then a bursting of the sphere when you chew through. This can lead to an explosion of taste impressions in the mouth. Spherification is based on the fact that sodium alginate, in the presence of calcium ions, can form a gel that is thermally stable. The calcium ions can come from, for example, calcium chloride or calcium lactate, or they can be naturally present in what you want to spherify.

Fig. 9.4 Pearls made by spherification of a solution of a special seaweed extract (alginate), here with mint

Normally, chefs prefer to use calcium lactate, possibly mixed with calcium gluconate, because dissimilar to calcium chloride, it does not affect the taste. In ordinary spherification, a certain amount (e.g., a drop) of a liquid (e.g., fruit juice) in which sodium alginate is dissolved is dripped with a syringe or carefully placed with a suitable spoon into a solution with calcium ions. The calcium ions cause alginate to form a gel from the outside of the drop, creating a sphere, i.e., a kind of solid spherical shell with a liquid interior. The strength of the gel depends on the concentration of calcium ions and also on the acidity and possible alcohol in the liquid. If the liquid is too acidic ($pH < 5$), e.g., apple or lemon juice, a gel is not formed, but instead alginic acid, which is poorly soluble and only helps to make the liquid more viscous. This effect can be partially counteracted by adding sodium citrate, which raises the pH.

To stop the gel formation before the interior of the drop becomes solid, the spheres must be immediately transferred to clean water to remove excess calcium ions. However, it can be difficult to control, and it is not possible to wash away all the calcium ions, so after some time calcium ions may diffuse into the interior of the sphere, which then becomes more solid. A similar problem arises if the substance or liquid that one wants to make into spheres already contains a certain amount of calcium ions, such as dairy products. Spheres formed by conventional spherification must therefore normally be prepared just before they are to be used.

Laverbread: Welsh 'Kaviar'

Laver is the English name for the red seaweed species *Porphyra umbilicalis*, which is found in many places on the coasts of Wales, Scotland, and England. According to traditional recipes, the fresh seaweed is boiled for 6 h with a little salt until the seaweed leaves are cooked. The liquid is drained, and what remains is a black, thick, spinach-like puree, which is called laverbread (cf. Fig. 9.5). The salty taste has notes of oysters or olives. The hot laverbread can be eaten spread on toast and optionally garnished with toasted bacon. A classic recipe for breakfast in Wales is laverbread, which is heated in the juices and fat from fried bacon. It is eaten with bacon, mushrooms, and eggs. Another classic recipe uses laverbread mixed with oatmeal to make a kind of flat cake, which is fried in fat and eaten for breakfast with toasted bacon and mussels. The dish can be seasoned with lemon juice and pepper.

Laverbread is considered a delicacy by older Welsh people, and it is not without humor that laverbread is called "caviar from Wales," which is attributed to the Welsh actor Richard Burton. However, it is likely that laverbread has been considered a kind of survival food for coastal populations during times of food shortage. In 2017, laverbread was granted protected designation of origin (PDO) status by the European Union. It is said that laverbread should not be chewed, but rather sucked, so that the taste of the sea develops in the mouth.

It is noteworthy that a traditional Chilean use of the microalgae *Nostoc sphaericum* has recently been marketed as "Andean caviar."

Fig. 9.5 "Welsh kaviar" (laverbread) is produced by long-term cooking of the red seaweed purple laver

Chapter 10
Milt: "White Roe" from Fish Sperm

Milt is the sperm and gonads of male fish, and although it is not white in all fish, it is sometimes called white roe because it is the male counterpart to the female roe. The milt from codfish is also called curl because it has a curly shape. In flatfish, salmon, and herring, for example, the milt is smoother. The volume of the fully developed sperm sac can be almost as large as the roe sac in the female. However, unlike fish eggs, sperm cells are microscopically small, about 10,000 times smaller. More than 10,000 sperm cells are often released for each fish egg during fertilization in the water. As an example, a large salmon can contain up to half a liter of milt.

In some food cultures, milt is considered a great delicacy on a par with roe, and cod milt was previously a common food during the cold season. Thus, the traditional serving of boiled cod, or salt cod, was a meal that included the muscle meat, roe, liver, and milt, cooked and served separately. Sometimes the boiled cod head also found its way to the plate.

The milt from other common edible fish, especially flatfish, was previously fried and eaten together with the fish, in exactly the same way as the roe sacs. Nowadays, the milt from neither small nor large fish is eaten, partly because consumers do not buy whole fish, but most often fillets, where the bones, head, and possibly the roe and milt have been cut off and discarded. To the extent that, for example, whole flatfish is prepared in the home kitchen, many people do not eat the milt, perhaps because they do not know that it can be eaten and is actually a tasty delicacy or because they do not like its soft texture.

The milt is also eaten from marine animals other than fish. We have already described sea urchins, where what is called the roe can actually be both the entire male and female gonads. In Japan, there is also a tradition of eating the milt of squid. In other simpler mollusks such as mussels, e.g., blue mussels and oysters, we do not distinguish between roe and milt either but eat it all together with other viscera.

O. G. Mouritsen, K. Styrbæk, *Roe and Roe Gastronomy*,
https://doi.org/10.1007/978-3-032-13142-3_10

10.1 Milt: A Great Delicacy Around the World

In many places in the world, the milt of fish is considered a great delicacy, especially in South Korean and Japanese cuisine (cf. Fig. 10.1). The Japanese particularly appreciate the milt from monkfish, salmon, sea bream, *fugu*, and not least cod. The milt from cod is called *shirako*, which in Japanese quite aptly means "white children." *Shirako* is eaten raw, lightly salted, possibly with a little soy sauce, or on top of boiled rice, but can also be fried with *tempura* batter. You can also find soups with *shirako*, which is steamed in *dashi* and *mirin* or used as a filling in *gunkan-sushi* (battleship sushi).

Fig. 10.1 Serving cod milt (sperm)

In Romania, there is a tradition of eating milt from freshwater fish such as carp, and the milt is deep-fried and called *lapti*, which is derived from the Latin term for entrails. In Russia, people eat deep-fried milt from especially herring and various lean fish. In traditional British cuisine, you can find milt from cod, lightly fried in butter and spread on bread.

Tuna milt is a particular specialty in Sicily, where it is considered a gourmet product in the same way as *bottarga*. Tuna milt is also highly venerated in Sardinia and other places along the Italian and Spanish coast, where fish is landed. Tuna milt is more reddish than, for example, cod milt. It can be eaten as it is but is usually prepared as a traditional product in dried form, called *lattume* or *figatello*. *Lattume* is often eaten grated over pasta dishes in the same way as *bottarga*. It is also common to cut the milt into cubes, roll them in flour, and then fry them in olive oil with onions. A Sardinian specialty is to grill the dried *lattume*.

Herring Milt in Norway

Herring milt has been in the spotlight in recent years, especially in Norway, which has one of the world's largest herring fisheries. Because the industrial exploitation of herring focuses on fillets, it is often the case that 50–60% of the fish remains unused as human food, including some milt. However, heads, bones, intestines, and thus the milt can be used as feed in aquaculture. It was common in Norway to eat herring milt in the past, but modern consumers are not aware of it. The catch of herring milt in Norway amounts to around 30,000 tons per year, and projects are being worked on to exploit this valuable and nutritious raw material as a dietary supplement for humans.

10.2 Texture and Taste of Milt

Milt has a velvety and soft texture, which some will find pleasant. Others may find it unpleasantly flabby, and it may be difficult for some to psychologically free themselves from the knowledge of what it is. When cooked, milt attains a firmer texture (cf. Fig. 10.2). Milt is creamy and not grainy like many types of roe, where the individual eggs can be felt in the mouth and between the teeth. The fine texture is due to the fact that the sperm cells, unlike fish eggs, are microscopically small. The taste is predominantly sweet and umami with a very mild marine aroma.

10.3 Handling Milt

Milt is mostly sold during the cold winter months of the year, because it spoils easily. Milt is best handled and stored in the intact sperm sac. You can then cook the whole sac or squeeze out the sperm by cutting a hole in the outer membrane. It is

Fig. 10.2 Spanish dish with tuna milt and herbs in vinaigrette

important to keep the milt cool. Before use, it is washed in cold water and placed in salt water. It can then be boiled or fried. If you have never tasted milt before, you can probably appreciate it best if it is breaded and deep-fried, especially with *panko* breading.

Steaming milt enhances its sweet taste, and using *dashi* or sake enhances the umami taste. When deep-frying, the outside of the milt can become crispy, providing an interesting textural contrast to the soft and creamy interior.

10.4 Ingredients in Milt

There is very limited knowledge about the ingredients in milt from different fish species. Overall, fresh milt contains about 80–90% water, 5–15% protein, 1–3% fat, and 1–2% minerals (ash). Although there are large variations, a pattern emerges where eggs contain both more fat and more protein than sperm. The fats in milt, like in fish roe, consist predominantly of polyunsaturated fatty acids, and the ω-3/ω-6 ratio is comparable, typically around 4–7. The energy content is around 400 kJ/100 g and thus less than for most types of fish roe. There are considerable amounts of vitamins B_1, B_2, B_6, and B_{12} as well as the minerals calcium and iron. All in all, fish milt is a fairly nutritious food but not on par with fish roe.

The cholesterol content in fish roe is typically in the range of 0.1–0.2%. In milt from fish, it is generally significantly higher with values up to 0.8%. For example, 0.61% is found for salmon and 0.71% for plaice. Biologists believe that the high cholesterol content is important for mechanically strengthening the survival of the small sperm cells on the way to fertilizing the eggs in the cold seawater.

Chapter 11
Salting, Drying, Smoking, Preserving, Freezing, and Maturing

In this chapter, we will look at how to loosen the eggs from the roe sac of fish and how to preserve and process the eggs afterward before use or during maturation and storage. We will also look at general treatments of whole roe sacs both for the purpose of preservation and for the production of special flavors and textures. The special treatment of sturgeon roe and caviar has been discussed in detail earlier (cf. Sect. 3.5).

11.1 How Do You Get the Eggs from the Roe Sac?

The release of the individual eggs from the roe sac proceeds differently depending on the size of the eggs. In all cases, the release is followed by a quick wash of the eggs in saltwater and removal of remnants of the connective tissue of the roe sac. This is immediately followed by salting, either in saltwater or by dry salting, and during the entire process, an attempt is made to keep the temperature as low as possible.

Larger eggs from, e.g., sturgeon and trout are traditionally separated from the roe sac industrially by pressing it through a stainless-steel mesh or sieve (cf. Fig. 3.5), which retains the connective tissue. In households, a whisk is often used to release the connective tissue. A quick and light poaching makes this job considerably easier but can lead to dull eggs and eggs with a rubbery consistency. Smaller eggs from, e.g., herring, capelin, and vendace are separated from the roe sac industrially in a rotating drum sieve.

It is possible to use special enzymes, e.g., collagenase and pepsin, which can break down proteins, to dissolve the connective tissue in the roe sac and thus facilitate the release of the individual eggs. In some cases, enzymes extracted from crab livers or fish intestines are used.

O. G. Mouritsen, K. Styrbæk, *Roe and Roe Gastronomy*,
https://doi.org/10.1007/978-3-032-13142-3_11

11.2 Industrial Products and Preservation

All commercial caviar products have undergone salting for reasons of taste, texture, and preservation. The salt content (NaCl) is typically 2.7–3.3%. In addition, certain types of kaviar have added substances that act to stabilize and preserve the product and prevent oxidation of fats in particular. The addition of acids regulates the pH to the range $5.1 < pH < 6.3$, which is important for preserving the quality of the roe during storage by preventing unwanted microbial activity of molds and yeasts and, for example, bacterial formation of biogenic amines that are hazardous to health. Finally, dyes and flavors may be added.

Although roe are an important product on the international market, there is surprisingly little information available on the chemical composition, food safety, and quality parameters of the products. In many cases, analyses have found small amounts of additives that are not even declared on the product, such as lactic acid (E 270) and acetic acid (E 260). These small amounts probably do not pose any health risk.

Commercial kaviar products are usually marketed in glass jars with a limited shelf life, and the products must be stored in the refrigerator. In Spain and Portugal, you can find canned roe products from, for example, sardines, squid, and mackerel (cf. Fig. 11.1). In the Scandinavian countries, canned or tubed lumpfish roe and cod roe products are particularly popular (cf. Sects. 5.4.2 and 5.4.3).

Fig. 11.1 Canned mackerel roe

11.3 Salted Roe

Salting is an essential part of the preparation and preservation of both caviar and all other types of kaviar from fish eggs in the form of loose eggs. In addition, salting also acts as a flavor enhancer and affects the structure of the eggs and thus the texture of the product. However, there is also a tradition in many food cultures to salt the whole roe sac for products where the cohesive eggs or the paste-like products that various cooking methods lead to are used. The salted roe products *tarako* and *mentaiko* are described under Cod (cf. Sect. 5.4.1).

11.4 Dried Roe

Drying fish roe has been a very common way of preserving roe in most fishing cultures, not least in warm climates. Heavy dry salting and subsequent drying and pressing is a traditional way of preparing fish roe in both Mediterranean countries, e.g., *bottarga* in Italy, *botargo* in Spain, *poutargue* in France, and *avgotaracho* in Greece, and in several places in Asia, e.g., *karasumi* in Japan. Along the Mediterranean coast and in Southern Spain and Portugal, a range of *bottarga* products are sold at public markets (cf. Fig. 11.2).

It is especially the roe of the flat-headed, gray mullet (*Mugil cephalus*) that have been used for *bottarga* and *karasumi* because of the roe's mild and only slightly bitter taste. However, roe from tuna and swordfish are also used for *bottarga* (cf. Fig. 11.3). *Bottarga* from tuna has a particularly strong taste and is firmer than the

Fig. 11.2 Dried gourmet roe products are sold at a market in Cádiz

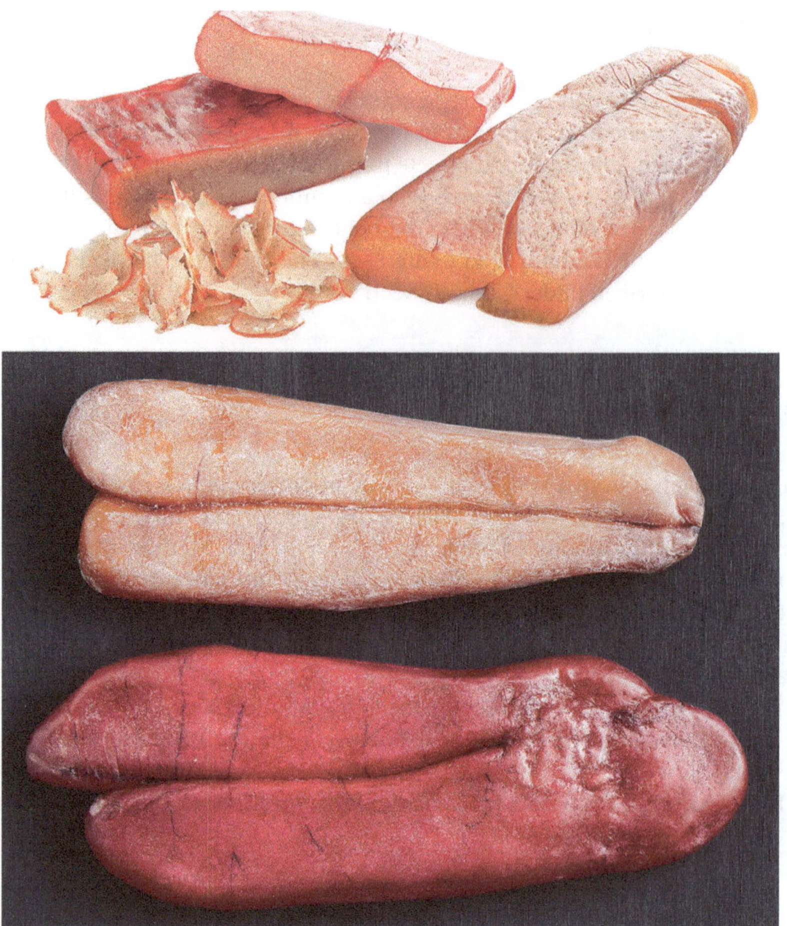

Fig. 11.3 Selection of *bottarga* from different types of fish roe

somewhat softer *bottarga* from mullet. In addition, the tuna product is very dark brown compared to the light and orange-colored mullet roe. Cod roe have also been used for *bottarga*, and because of the much lower fat content and especially waxy fats in cod roe, *bottarga* from cod is softer and has less of a characteristic fishy taste from oxidized fats.

In Europe, the tradition of *bottarga* can be traced back to the Phoenicians (1,100–200 BC), but perhaps the starting point was Egypt, which is suggested by the fact that the word *bottarga* is derived from the Arabic term *butarhah*, which originates from Coptic. There are also sources that describe that *outarakhon* (which means pickled eggs) were widespread in the Greek colony of Byzantium around 600 BC. *Bottarga* was for centuries a valuable trade item in the Mediterranean area. Today, *bottarga* is considered a luxury product, and in recent years, chefs around the world have become aware of *bottarga*'s ability to elicit umami and give a certain "Mediterranean flavor" to food.

Bottarga is still a great delicacy, for example, in Sicily and Sardinia (*butàriga*), and you can find it in Spanish tapas bars, where it is eaten with olive oil or a little lemon juice. In Italy, *bottarga* is often used finely shaved and sprinkled or grated onto a pasta dish in the same way that Parmesan cheese is used to add salt and umami. Slices of *bottarga* can also be lightly toasted.

Sardinia is the main producer of *bottarga*, where the product has the status of a recognized traditional, regional product, Prodotti Tipici Sardi, but *bottarga* is also produced in a number of other countries, such as the USA and Norway, where cod roe are used in the latter country. Similar products are also found in Japan, South Korea, and Taiwan. In South Korea, both mullet and drum roe (from the Sciaenidae family) are used, and the roe are marinated in soy sauce and then semi-dried in sunlight. In Taiwan, the membrane is removed from the dried and salted roe, soaked in sake, and then roasted over an open fire for a few minutes until the surface becomes crispy and milky white. It is then eaten thinly sliced with Chinese radish and spring onions.

In Japan, dried mullet roe in the form of *karasumi* are one of the three most prominent marine delicacies (*tenka no sandai chinmi*) along with sea urchin roe (*uni*) and fermented and marinated sea cucumber entrails (*konowata*). Japanese *karasumi* is less dried than *bottarga* and therefore has a slightly softer texture. Traditionally, sake is drunk with *karasumi*, which is not a cheap pleasure, and it is used both as an appetizer and as part of traditional *kaiseki* dishes. Interestingly, *karasumi* entered Japanese cuisine with traders via the Silk Roads from Greece and Egypt in the 1650s, probably first to China and from there to Nagasaki, which was the gateway to Japan from the outside via the small island of Dejima in Nagasaki Bay. Here, mackerel roe and sea salt from the Nagasaki area were first used, and only later mullet roe, which in its migration reaches the Nishi Peninsula near Nagasaki. Commercial Japanese *karasumi* production is still associated with Nagasaki and sea salt from this area. Elsewhere in Japan, *karasumi* is produced from cod, tuna, and *hamachi*.

Bottarga in Pepys' Diary from 1661

Bottarga is mentioned in the famous British author Samuel Pepys' diary, which was written in London in the 1660s. On a warm evening on June 5th, Samuel Pepys is sitting in the garden drinking wine and eating *bottarga*. *Bottarga* stimulates thirst, so it is not without reason that Pepys notes in his notes that he does not go to bed until after midnight and that he is then very drunk.

11.4.1 Bottarga *Production*

Bottarga is made from the whole roe sacs, which are taken from tuna, swordfish, mullet, and various fish from the cod family. The production consists of two stages: salting and then drying. First, the roe are massaged by hand to remove air pockets that would otherwise later contribute to oxidation and rancidity of the oils in the fish eggs. The roe are then dry salted by placing it in sea salt for a few weeks, after which it is hung to air-dry for about a month. During this time, the roe become darker in color. The salt draws liquid from the roe, which then become firm and hard and especially waxy if the *bottarga* is made from mullet roe, which has a high content of waxy fats. Other recipes specifically for mullet roe involve a shorter salting process, 4–5 h, after which the whole roe sacs are compressed and dried until they have lost 30% of their weight.

During the production of *bottarga*, the eggs fuse into a solid mass, and since this works best with small eggs, roe from fish with large eggs, such as salmon and trout, are not used. The quality is ultimately judged by an inspector's feeling for the firmness of the product.

Due to the high fat content, drying the roe takes a long time, typically 3–6 weeks. The final product contains 50–60% fat and 20–40% protein. The quality of the polyunsaturated fatty acids does not seem to be significantly affected by this rather extensive process. However, the preparation time depends on local traditions and is conditioned by how firm the final product is desired. In some cases, the surface is protected with a layer of melted beeswax, which must be removed before the roe are eaten.

Since *bottarga* is made from whole roe sacs, the membrane that sits on the outside is still on the product after the drying process, but it has become a bit like dry paper. You can eat this dry layer, but it is best to remove it, as it otherwise leaves some tough remnants and can be bitter.

During preservation and subsequent storage, the roe develop a strong umami taste, because free amino acids, especially glutamate, and certain free nucleotides are produced, which enhance the umami taste (cf. Table 5 at the end of the book). It is during the last part of the drying process, which takes place in the sun, that the umami taste is developed, and the final product contains about four times as much glutamate as fresh mullet roe. This is because the roe still contains active enzymes, which are also effective when a lot of salt is present. In addition, sugars combine with proteins in slow Maillard reactions, which give the roe a brownish color and the many delicate flavors and aromas that characterize browning reactions. The resulting product is therefore salty, has umami taste, and also develops some of the nutty notes that characterize mature hard cheeses like Parmesan cheese. The dried roe are firm and slightly waxy and are typically used thinly sliced or, as previously mentioned, grated over or in a dish, e.g., pasta. It can also be used as a snack. *Bottarga* can also be made from cephalopod roe, e.g., from octopus (cf. Fig. 7.2), but it is not commercially available but can be found in a few places in Spain and Portugal.

The Umami Potential Grows Along the Way
In the process from the fresh mullet roe to the finished *bottarga*, the contents of the substances that provide umami taste and umami synergy grow (cf. Tables 2 and 5 at the end of the book). The content of free glutamate grows from 38 mg/100 g to 158 mg/100 g, and it is especially drying in the sun that has the greatest effect. At the same time, the content of free nucleotides grows, for example, IMP goes from 5 to 12 mg/100 g and GMP from 3 to 7 mg/100 g. Expressed by the equivalent umami concentration (EUC, cf. Table 4 at the end of the book), the growth is from 589 mg/100 g to 5565 mg/100 g from the fresh roe to the finished *bottarga* product, meaning that there is almost ten times more umami intensity in *bottarga* than in the fresh roe.

The dried roe have been called poor man's caviar. It is available in the trade as whole roe sacs, often in pairs from the same fish. It is typically eaten finely sliced or grated like Parmesan cheese. It is also sold as a paste consisting of grated *bottarga*. Nowadays, it is mainly southern Italy, Sicily, and Sardinia that keep the tradition of *bottarga* production alive as a regional specialty, and, besides Italy, Spain and France are the largest consumers.

Bottarga can easily be kept on hand to grate over a pasta dish, boiled rice, mashed potatoes, risotto, omelette, or a simple green salad which need some umami taste (cf. Fig. 11.4). *Bottarga* cannot withstand heating, because it loses its delicate flavor and aroma.

11.4.2 Karasumi*: Salted, Dried, and Fermented Roe*

There are preparations of roe in Asian cuisines related to *bottarga*, which is not surprising, since this way of preparing fish roe came to the East from the West via the Silk Roads. Japanese *karasumi* is usually made from mullet roe in an extensive process that includes salting, pressing, drying, and maturing—a process that takes up to 40 days under varying temperature conditions. The process is said to have been invented by Yusuke Takano in 1675.

After the whole roe sacs have been removed from the fish, they are gently cleaned in water, smeared with sea salt, and then left in a barrel for 3–6 days. They are then partially desalted in a tub of fresh water for about a day until they have achieved the appropriate softness. The salt content at this point is crucial for the taste and aroma of the finished product. The roe sacs are then pressed between some wooden boards. This is followed by about 10 days of alternating pressing at night and sun drying during the day. The final product (cf. Fig. 11.5) is wiped to remove any oil that may have been squeezed out along the way.

Fig. 11.4 *Bottarga* from mullet grated over asparagus broccoli

During maturation, a slight lactic acid fermentation takes place, which contributes to the deep umami taste and long aftertaste that is known from hard and aged cheeses, e.g., Parmesan cheese. The salt content of the finished product is nowadays around 4–5%, whereas it was previously somewhat higher (7–8%). The water content is as low as 20–25%. Like *bottarga*, *karasumi* is used finely grated over other

Fig. 11.5 *Karasumi,* a dried Japanese roe product, here made from mullet roe

dishes, but it can also appear as a side dish grated and mixed with rice wine vinegar. The Japanese use both *karasumi* and *tarako* mixed in pasta dishes (*pasta tarako*) (cf. Fig. 5.20). In modern Japan, *karasumi* is also eaten as a "high-class" side dish for drinks.

11.5 Cooked Roe

Boiling, steaming, or other heat treatments of roe are mainly used for smaller fish species, where the roe either consists of very small eggs, where the roe sacs are difficult to remove whole, or where the food tradition is to cook whole fish that are in the season when they carry roe. This is especially the case for flatfish (plaice, turbot, halibut, flounder, dab, halibut, and sole) or round fish such as mackerel, cod, and herring. Unfortunately, many consumers are not used to eating the roe from these fish.

Fresh cod roe should be cooked immediately (cf. Fig. 11.6). Small cod roe sacs are cooked for about 3 min and very large ones up to 10 min. If the roe are to be eaten immediately after cooking, it should be left in the boiling water for 15 min. Cooked cod roe will keep for a few days in the refrigerator.

Recipe 11.1 Freshly Cooked Cod Roe with Lime Cucumber and Capers (Fig. 11.6)
Serves 8–10

700 g whole fresh cod roe sacs
Zest and juice of 1 unsprayed lemon
Salt
Court bouillon
1 l water
30 g salt
½ tsp peppercorns per l water
3 tbsp apple cider vinegar
1 onion
Parsley stalks
2 bay leaves

Lime, cucumber, and capers

- *Condiments for 4*
- 1 firm cucumber
- 1 unsprayed lime
- 3 tbsp capers
- 1 tsp sugar
- A little salt
- 4 tbsp crème fraiche 18%
- ½ handful chopped parsley

1. Place the cod roe sacs on a piece of cling film, grate the lemon zest over it, and squeeze a little lemon juice on it. Season with salt.
2. Roll cling film tightly around the roe and twist the ends together.
3. Put the roe in a saucepan, cover with water, and add the remaining ingredients for the court bouillon. Bring the water to a boil, let the roe simmer under a lid over low heat for 25–30 min, turn off the heat, and let the roll cool in the broth for the same amount of time.
4. Peel the cucumber, split it lengthwise, remove the seeds, cut it into cubes, and place the cubes in a small bowl.
5. Grate the lime zest over it, drizzle with the juice, season with a little salt and sugar, stir in the capers, crème fraiche and chopped parsley, and season again.
6. Remove the film from the roe and serve the roe immediately or refrigerate.
7. Cut the roe into slices and serve with wholemeal bread, a spoonful of lime pickle, and, if desired, a few lemon wedges.

Fig. 11.6 Freshly boiled cod roe with lime cucumber and capers

11.6 Smoked Roe

All roe can be smoked, both whole roe sacs and loose eggs from fish with large eggs such as trout and salmon. The best known is probably smoked cod roe (cf. Fig. 11.7), which is a sought-after delicacy in season. Both hot-smoked and cold-smoked cod roe are produced.

Fig. 11.7 Smoked cod roe

Cod roe can be cold smoked in whole roe sacs, which have been carefully cleaned in cold water so that the thin membrane does not break. The roe sacs are then dry salted for 6–8 h. The excess salt is subsequently washed off in clean water, and the roe are blanched for a few minutes in boiling water, whereby the roe swell slightly but is still soft and not cooked all the way through. The whole roe sacs are then laid out on a grid and placed in a smoke oven at 30–32 °C for 8 h. The smoked cod roe are suitable in *taramasalata* together with a green salad or sliced on open sandwiches. Mullet roe are also suitable for smoking, and this is done in the same way as cod roe.

Loose eggs from salmon roe can also be cold smoked, which is done slowly at temperatures just above freezing to a point where the smoke has given the eggs a note of smoke without overpowering the roe's own flavor. Smoked salmon eggs can contribute to the flavor of green salads, fish or shellfish dishes, or as a decoration on canapés. Smoked salmon roe are available as a commercial product.

The Smoke Master's Hot-Smoked Cod Roe

Poul Rasmussen has worked in the fishing industry for 50 years, and he has smoked fish and roe for at least 40 years. As is customary on the island Funen in Denmark, all popular people have an affectionate nickname, and Poul is of course called Poul Fisk or Fiskerpoul, which suits him well. With a training as a laboratory technician, Fiskerpoul has worked with fish and smoking in many towns in Denmark over the years, and he was director of the wholesale company Albanigades Fisk og Vildt until his retirement a few years ago.

Fiskerpoul is *the* expert and has built cold and hot ovens for smoking both in Denmark and Spain. Here is his recipe for his hot-smoked cod roe, but, as he emphasizes, there are almost as many different recipes as there are smoke masters.

The fresh roe sacs are wet salted (8–10%) for 20–25 min. The roe are now placed on trays in a hot smoker, right after it has been used to smoke fish. The oven is turned off, and the roe now stands in the cooling oven overnight, where the residual heat works to slowly dry the roe. In the hot oven, the roe initially swells and becomes firmer. The oven is kept open so that the moisture can escape while the roe cools. The next day, the oven is heated to 76–78 °C, which is determined for bacteriological reasons. When the oven is hot, beech sawdust is placed on the heating elements, which does not ignite the sawdust but just causes it to develop smoke. After an hour or so, the roe have become golden brown and has acquired a suitable firmness, so that it is cut-resistant.

The smoked roe can be kept for about 8 days in the refrigerator. You can smoke other types of roe according to the same recipe, including loose roe such as lumpfish roe, which is then smoked in stockings.

Fiskerpoul says that some special customers want cod roe soft and slightly runny in the middle. The recipe for this is as above, except that the roe are not dried in the hot oven but at a lower temperature. The slightly moister roe, together with the smoke, gives the soft cod roe a more bitter taste, which some prefer.

11.7 Storing and Freezing Roe

Due to the content of unsaturated fats, fresh and unsalted roe have a very short shelf life. It should therefore be stored in a cool, dark place in the refrigerator if not frozen. It is best to eat or prepare (boil, fry, and smoke) the fresh roe immediately. Fresh roe should therefore be eaten immediately and preferably within a few days of being stored in the refrigerator. For example, fresh, unsalted lumpfish roe can be kept in the refrigerator for 3–4 days.

Roe sacs and loose eggs (kaviar) can be frozen. Whole roe sacs must be handled carefully so that the fragile membrane around the roe is not damaged. Roe with small eggs tolerate freezing better than roe with large eggs. Freezing should take place while the roe are completely fresh and before it has been exposed to heat or sunlight. Before freezing, whole roe sacs must be carefully cleaned, and blood and all remains of the fish's intestines must be removed. It is advantageous to lightly salt the roe before freezing to protect the eggs from freezing damage. The cleaned roe sacs are pricked in a few places with a sterile needle, after which they are packed in plastic bags, which are lightly pressed together to remove as much air as possible.

The shelf life of frozen roe is best for the types with the lowest fat content, but since all types of roe contain relatively high amounts of unsaturated fat, the shelf life in the freezer is no more than 3 months. Freezing at −20 °C for 1–3 months and thawing for a few hours at room temperature do not seem to affect the quality of the roe. Any differences are no greater than typical and natural variations of the product.

11.8 Parasites and Toxins in Fish Roe

Like other marine products, roe can be contaminated with unwanted parasites, microorganisms, or toxins that are either environmental toxins or substances that are developed during microbial decomposition of the roe, i.e., when the roe are not fresh or have not been properly treated.

Unwanted parasites in roe are usually not a problem in roe from aquaculture, but in some cases roe from wild fish can be infected with flukes, tapeworms, or roundworms. If you are not sure about the quality of the roe, the best way to be sure is to freeze it at −18 °C for at least 24 h or 72 h at −10 °C.

Roe from completely fresh fish is microbially sterile but can become infected by bacteria (e.g., *Pseudomonas* sp., *E. coli.* bacteria, and a wide range of other bacteria) during processing, and parasites from the liver and muscle can move into the roe. If the roe are not treated properly, in addition to smelly breakdown products of fats and proteins, so-called biogenic amines can be formed, such as histamine, which is formed from the amino acid histidine. Histamine poisoning is characterized by a burning sensation in the mouth and throat, skin reactions with redness and tingling, nausea, vomiting, and headaches. This type of microbial degradation can be limited by the use of preservatives or pasteurization of the roe.

Recipe 11.2 Fresh Roe Sacs in Crispy Dough with Three Types of Dressing (Fig. 11.8)
Serves 4

A selection of different fish roe in their own roe sac
1 lemon

Dough
40 g cornstarch
40 g gluten-rich flour, e.g., Manitoba
2 pinches baking powder
1–1½ dl cold sparkling water
½ dl vodka or other pure alcohol (makes the crust crispy)
1 tsp neutral oil
A little fine salt
1–2 dl oil for frying

First Dressing
1 dl lime juice
½ dl sugar
2 tbsp fish sauce
2 tbsp soy sauce
A little crushed garlic and chili

Second Dressing
2 dl crème fraiche 18%
2 tsp fish mustard stirred in a little cold water
Fresh, finely chopped chives
Salt and freshly ground pepper

Third Dressing
2 dl soy sauce
Wasabi powder or paste
½ finely chopped spring onion

1. Mix the dry ingredients for the dough, add the liquid, and stir carefully.
2. Refrigerate the dough until ready to use.
3. Place the bowl in a slightly larger bowl with water and ice cubes so that the dough also stays cool while you prepare it.
4. Heat the oil to 165 °C.
5. Dip the roe sacs in the batter, place a few pieces at a time in the oil, fry until crispy, and place on absorbent paper. If you want a thicker crust, sift flour over the roe sacs before dipping them in the batter.
6. First dressing: Mix everything together and pour into a bowl.
7. Second dressing: Stir the ingredients together, season to taste, and pour into a bowl.
8. Third dressing: Stir together soy sauce and *wasabi* (the dressing should be slightly spicy) and pour into a bowl with spring onions.

Place the crispy roe sacs on a platter to share and serve with the dressings on the side and a few lemon wedges. It is easiest to use smaller roe sacs for this dish because the roe tend to be runny for larger ones.

Fig. 11.8 Fried roe sacs in crispy batter with three types of dressing

The Technical Details

Tables

The tables are adapted from Mouritsen and Styrbæk (2023) and Schmidt et al. (2024).

Table 1. Proximate Composition and Egg Sizes

Proximate composition and egg size of roe from a series of fish, crustaceans, mussels, echinoderms, and cephalopods as derived from the literature. All data refer to wet weight, and, in some cases, the data from the literature have been renormalized from dry weight to wet weight for the sake of comparison. Water content is given in %, and in case the water content has not been reported in the literature, it has been estimated (in which case the value is given in parenthesis). Contents of protein, lipid (fat), carbohydrate, and cholesterol are given in mg/100 g wet weight. ω-3/ω-6 denotes the ratio between total amount of ω-3 and ω-6 fatty acids. The egg size is given as an approximate average diameter (*d*) in mm since many egg types are not spherical in shape. Data for diameters for sturgeon eggs (denoted by *) are quoted from the review by Siddique et al. (2014). "-" denotes that data is not available. In several cases, there is no full accordance between data from different sources. Variations in data can also be caused by differences in degree of maturation and harvesting time of the eggs, the feeding status of the animal in question, and whether the data is from wild or farmed animals. Some data is from whole gonads at different stages of maturation. The data reported in the table all derive from the international scientific literature and may in some cases deviate from data (e.g., for egg diameters) reported by companies for commercial products that often are optimized with respect to certain gastronomic parameters like color, egg size, and texture. Concerning the instrumental uncertainty of the individual sets of data, which can vary greatly, the reader is referred to the literature. Other inaccuracies encountered

O. G. Mouritsen, K. Styrbæk, *Roe and Roe Gastronomy*,
https://doi.org/10.1007/978-3-032-13142-3

in the literature data are caused by a lack of precise description of the species investigated, in some cases even a clarification of the genus. Some of these data are included for completeness and illustration of the situation in quantitative characterization of roe. Carb = carbohydrate; Chol = cholesterol

Fish									
Species	Scientific name	Water	Protein	Lipid	Carb	Chol	ω-3/ω-6	*d*	References
Alaska pollock	*Theragra chalcogramma*	73.3	19.2	2.8	-	-	12.8	-	Vasconi et al. (2020)
	—	67–72	26	5	-	0.32	-	1.3–1.5	Bledsoe et al. (2003)
	Tarako (salted)	-	-	3.7	-	0.35	19.0	-	Tocher and Sargent (1984)
Ayu sweetfish	*Plecoglossus altivelis*	60.3	25.9	10.6	-	-	-	-	Suzuki and Suyama (1983)
Bonito	*Euthynnus affinis*	73.0	18.2	4.3	2.76	0.094	5.3	-	Intarasirisawat et al. (2011)
Brill	*Scophthalmus rhombus*	(91)	-	0.7	-	0.11	7.9	1.42	Hachero-Cruzado et al. (2011)
	—	91	8.4	1.2	-	-	8.9	-	Hachero-Cruzado et al. (2013)
Bullrout	*Myoxocephalus scorpius*	-	-	-	-	-	-	2.3	Hoffmann (2003)
Burbot	*Lota lota*	-	-	7.0	-	0.05	4.7	-	Kaitaranta and Ackman (1981); Kaitaranta and Linko (1984)
	—	77.7	13.2	6.8	-	-	-	-	Vúorela et al. (1979)
	—	64.5–77.7	13–16	6.8–9.4	-	-	-	-	Krzynowek and Murphy (1987)
	—	-	-	6–7	-	-	-	-	Bledsoe (2006)
	—	-	-	-	-	-	-	1.2–1.4	Cohen et al. (1990)
Carp	*Cyprinus carpio carpio*	68	18.3	10.8	-	0.011	-	-	İnanlı et al. (2019)
	—	68.7	22.8	6.3	-	-	-	-	Suzuki and Suyama (1983)
	Cyprinus sp.	70	24	2	-	-	-	0.8–1.6	Bledsoe et al. (2003)

(continued)

Fish									
Species	Scientific name	Water	Protein	Lipid	Carb	Chol	ω-3/ω-6	*d*	References
Catfish	*Ictalurus punctatus*	64.5	24.6	8	-	0.64	1.1	-	Bledsoe et al. (2003); Bledsoe and Basco (2006)
Cod	*Gadus morhua* (Atlantic)	71.6	19.6	3.2	-	-	13.2	-	Vasconi et al. (2020)
	—	74	-	3.4	-	0.21	15.4	1.3–1.4	Tocher and Sargent (1984)
	—	-	-	-	-	-	10.7	-	Salze et al. (2005)
	—	92	-	0.77	-	0.09	17	-	Fraser et al. (1988)
	—	67.4	25.8	5.2	-	0.32	-	-	Iwasaki and Harada (1985)
	—	70.0	24.3	1.7	3	-	-	-	Krzynowek and Murphy (1987)
	—	-	-	-	-	-	-	1.5	Hoffmann (2003)
	—	78–80	16–20	0.3–0.7	-	-	10.3	1.3–1.4	Bledsoe et al. (2003); Bledsoe and Basco (2006)
	Gadus macrocephalus (Pacific)	67.9	26.5	4.3	-	0.30	-	-	Iwasaki and Harada (1985)
Flatfish	*Limanda herzenstein*	68.4	27.2	3.3	-	0.41	-	-	Bledsoe et al. (2003)
	Microstomus achne	68.3	27.1	3.5	-	0.42	-	-	Bledsoe et al. (2003)
Flounders	*Hippoglossoides dubius*	70.4	24.0	3.7	-	-	-	-	Suzuki and Suyama (1983)
	Paralichthys olivaceus	70.2	21.2	7.3	-	0.54	-	0.8–1.2	Bledsoe et al. (2003)
	Platichthys flesus	-	-	-	-	-	-	1.1	Hofman (2003)
Flying fish	*Cheilopogon furcatus*	70.3	11.3	2.9	-	-	5.6	-	Lee et al. (1992)
	—	-	-	-	-	-	-	<2	Bledsoe et al. (2003)
	Cypselurus opisthopus hiraii	80.9	14.1	3.1	-	-	-	-	Suzuki and Suyama (1983)
	Tobiko (salted)	-	-	3.2	-	0.43	9.6	-	Tocher and Sargent (1984)
Garfish	*Belone belone*	-	-	-	-	-	-	3.0–3.5	Ehrenbaum et al. (1904)

Greater weever	*Trachinus draco*	-	-	-	-	-	-	1.0	Hoffmann (2003)
Halibut	*Hippoglossus hippoglossus*	(90)	5.9	-	-	-	-	-	Rønnestad et al. (1993)
	—	(90)	-	1.4	-	-	-	-	Evans et al. (1996)
	—	90	-	-	-	-	-	3.0–3.1	Riis-Vestergaard (1982)
	—	(90)	-	1.2	-	0.12	7	-	Falk-Petersen et al. (1986)
Haddock	*Melanogrammus aeglefinus*	86	-	2.8	-	0.27	8.7	1.2–1.4	Tocher and Sargent (1984)
	—	-	-	9–10	-	-	8.8	-	Bledsoe and Basco (2006)
Hake	*Merluccius hubbsi*	-	-	6.6	-	0.38	5–9	-	Méndez et al. (1992)
	—	67	-	5–8	-	-	-	-	Bledsoe et al. (2003); Bledsoe and Basco (2006)
	Merluccius merluccius	(67)	16.5	11	-	-	-	-	Domínguez-Petit et al. (2010)
	—	-	-	-	-	-	-	3	Ehrenbaum and von Fischen (1905-1909)

(continued)

Fish									
Species	Scientific name	Water	Protein	Lipid	Carb	Chol	ω-3/ω-6	*d*	References
Herring	*Clupea harengus* (Atlantic)	-	-	2.0–2.7	-	0.05	8.3	-	Kaitaranta and Ackman (1981); Kaitaranta and Linko (1984)
	—	78.4	17.8	2.6	-	-	-	-	Vúorela et al. (1979)
	—	74	-	3.8	-	0.32	29.4	1.3–1.5	Tocher and Sargent (1984)
	—	75–82	13–17	3.0	-	-	-	-	Krzynowek and Murphy (1987)
	—	-	-	-	-	-	-	1.4	Hoffmann (2003)
	—	77	18.7	3.0	-	0.31	10–15	0.9–1.2	Bledsoe et al. (2003); Bledsoe and Basco (2006)
	—	80.1	8.7	4.5	-	0.05	3.7	-	Kalogeropoulos et al. (2012)
	Clupea pallasii (Pacific)	-	-	3.1–3.5	-	-	21.7	-	Kaitaranta and Ackman (1981)
	—	65.4	10.6	2.3	-	0.31	-	-	Lee et al. (2011); Tocher and Sargent (1984)
	Kazunoko (salted)	-	-	3.0		-	20.1	-	
Mackerel	*Scomber japonicus*	66.6	25.3	6.8	-	0.42	-	-	Iwasaki and Harada (1985)
	Scomber australasicus	-	-	7–8	-	0.34	9.4	0.8–1.0	Bledsoe et al. (2003); Bledsoe and Basco (2006)
Monkfish	*Lophius piscatorius*	82.1	11.5	5.3	-	0.31	-	-	Bledsoe et al. (2003)
	—	-	-	-	-	-	-	2.0–3.1	Bowman (1920)
Mullet	*Mugil cephalus*	61.5	22.6	13.7	-	0.44	2–3	-	Lu et al. (1979)
	—	50.4	28.7	19.8	-	0.49	-	-	Iwasaki and Harada (1985)
	—	50–62	19–29	9–20	-	-	-	-	Bledsoe et al. (2003)
	—	-	-	-	-	-	-	0.6	Lee et al. (1992)

Lumpfish	*Cyclopterus lumpus*	79.7	10.8	4.3	-	-	30.6	-	Vasconi et al. (2020)
	—	77–81	15.5	5.5–6.1	-	-	29.5	-	Basby et al. (1998)
	—	-	-	-	-	-	-	2.2	Hoffmann (2003)
	—	-	-	-	-	-	-	2–5	Bledsoe et al. (2003)
	—	69.8	16.8	4.2	-		7.3	-	Kalogeropoulos et al. (2012)
Perch	*Perca fluviatilis*	-	-	4.1	-	0.02	4.9	-	Kaitaranta and Ackman (1981); Kaitaranta and Linko (1984)
	—	84.9	8.2	3.9	-	-	-	-	Vúorela et al. (1979)
	—	85	-	-	-	-	-	-	Bledsoe et al. (2003); Bledsoe and Basco (2006)
	—	-	-	-	-	-	-	2.0–2.5	Muus and Dahlstrøm (2017)
Pike	*Esox lucius*	64.0	19.4	12.7	-	-	4.3	-	Vasconi et al. (2020)
	—	64–67	14–27	1.5–2.4	-	-	-	2.5–2.8	Bledsoe et al. (2003)
Plaice	*Pleuronectes platessa*	-	-	-	-	-	-	1.6	Hoffmann (2003)
	—	-	-	1.8	-	0.31	-	-	White and Fletcher (1987)
Pollock	*Pollachius virens*	72	-	4.0		0.44	-	0.9–1.1	Tocher and Sargent (1984)
Roach	*Rutilus rutilus*	-	-	3.7	-	0.23	11.8	-	Kaitaranta and Ackman (1981); Kaitaranta and Linko (1984)
	—	72.6	19.0	4.2	-	-	-	-	Vúorela et al. (1979)
	—	66–67	24–26	2–4	-	-	-	-	Bledsoe et al. (2003); Bledsoe and Basco (2006)
	—	-	-	-	-	-	-	1	Muus and Dahlstrøm (2017)

(continued)

Fish									
Species	Scientific name	Water	Protein	Lipid	Carb	Chol	ω-3/ω-6	*d*	References
Salmon	*Oncorhynchus keta* (chum)	53.5	29.6	12.8	-	-	15.1	-	Vasconi et al. (2020)
	—	57.6	27.0	14.1	-	0.45	-	-	Iwasaki and Harada (1985)
	—	50–56	27–35	12–20	-	0.45	-	4–5	Bledsoe et al. (2003)
	Oncorhynchus kisutch (coho)	-	-	-	-	-	-	3.5–4	Bledsoe et al. (2003)
	Oncorhynchus nerka (sockeye)	56–58	20–29	10–13	-	-	-	4–4.5	Bledsoe et al. (2003)
	Oncorhynchus gorbuscha (pink)	50–60	23–38	11–13	-	-	-	3.5–5	Bledsoe et al. (2003)
	Oncorhynchus tshawytscha (chinook)	55.6	26.2	10.6	-	-	12.4	4.2–8.5	Bekhit et al. (2009)
	—	51–70	21–34	8–18	-	-	-	6–7	Bledsoe et al. (2003)
	Salmo salar (Atlantic)	45	32.7	8.9	-	0.26	7.0	-	Kalogeropoulos et al. (2012)
	S. salar (Pacific)	60.6	24.5	5.5	-	-	11.2	-	Shuang et al. (2019)
	—	-	-	-	-	-	10.7	-	Ma et al. (2020)
	Unspecified	45.1	26.8	12.7	8.3	-	8.1	-	Mol and Turan (2008)
	Ikura (salted)	-	-	14.5	-	0.59	18.0	-	Shirai et al. (2006)
	—	43–51	30–32	6–9	-	-	-	-	Hayashi et al. (1990)
Sand eel	*Ammodytes lancea*	64		5.1	-	0.24	11.6	0.3	Tocher and Sargent (1984)
Sander	*Sander* sp.	60–71	18–26	1–11	-	-	-	-	Bledsoe et al. (2003)
	—	-	-	-	-	-	-	2.0–2.5	Muus and Dahlstrøm (2017)
Sardine	*Sardinops melanosticta*	68.7	24.4	6.0	-	0.40	-	-	Bledsoe et al. (2003)
Seabream, red	*Pagrus major*	73.4	20.3	4.9		0.37	-	-	Bledsoe et al. (2003)
Saury	*Cololabis saira*	74.9	19.8	3.8	-	-	-	-	Suzuki and Suyama (1983)
	—	-	-	11	-	-	8–10	0.9–1.1	Bledsoe et al. (2003)

Skipjack tuna	*Katsuwonus pelamis*	72	20.2	3.4	2.35	0.17	5.0	-	Intarasirisawat et al. (2011)
Smelt	*Mallotus villosus*	81.7	8.1	4.5	-	-	17.4	-	Vasconi et al. (2020)
	—	80.4	10.8	4.3	-	-	14.7	-	Lee et al. (2011)
	—	70	-	6.8	-	0.21	24.8	1.0–1.2	Tocher and Sargent (1984)
	—	-	-	18–19	-	-	11	-	Bledsoe and Basco (2006)
	Spirinchus lanceolatus	61.4	24.1	13.2	-	0.56	-	-	Iwasaki and Harada (1985)
Sole	*Solea solea*	(90)	-	1.8	-	-	4	-	Parma et al. (2015)
	—	-	-	-	-	-	-	1.4	Hoffmann (2003)
Sprat	*Sprattus sprattus*	-	-	-	-	-	-	0.8–1.5	Muus and Nielsen (2017)
Sturgeon	Paddlefish (*Polyodon spathula*)	-	-	-	-	-	1.5–2.8	2.2–3.2*	Wirth et al. (2002)
	Amur sturgeon (*Acipenser schrenckii*)	48.7	24.3	16.0	-	-	0.9	-	Gong et al. (2003)
	—	50.7	28.2	16.6	-	-	6.7	-	Shuang et al. (2019)
	—	-	-	-	-	-	3.7	-	Ma et al. (2020)
	—	-	-	-	-	-	-	2.9–3.3	Koshelev (2013)
	Beluga (*Huso huso*)	-	-	-	-	-	1.8–3.6	2.8–4.5*	Wirth et al. (2002)
	—	64.1	15.1	14.9	-	-	1.20	-	Ovissipour and Rasco (2011)
	—	56.2	25.4	14.8	-	-	-	-	Hamez et al. (2015)
	—	48.4	24.7	15.9	-	-	3.1	-	Mol and Turan (2008)
	—	57–77	17–32	11–18	6.9	-	-	-	Bledsoe et al. (2003)
	White (*Acipenser transmontanus*)	57.1	24.9	16.1	-	-	1.2	3.5–4.0*	Lopez et al. (2020)
	—	54.4	25	11	-	-	3.2	-	Caprino et al. (2008)

(continued)

Fish									
Species	Scientific name	Water	Protein	Lipid	Carb	Chol	ω-3/ω-6	*d*	References
	—	-	-	-	-	-	-	2.2–2.5	Bledsoe and Basco (2006)
	Hybrid (*Huso dauricus* × *A. schrenckii*)	47.7	25.6	16.2	-	-	0.94	-	Gong et al. (2003)
	—	-	-	-	-	-	-	(3.0–3.2)	Unpublished
	Kaluga (*H. dauricus*)	50.5	28.1	15.1	-	-	6.5	2.5–4.0*	Shuang et al. (2019)
	—	-	-	-	-	-	5.4	-	Ma et al. (2020)
	Persian (*Acipenser persicus*)	51.5	24.2	14.7	-	-	3.6	-	Mol and Turan (2008)
	—	-	-	-	5.4	-	-	3.3–3.9	Nazari et al. (2009)
	Russian (*Acipenser gueldenstaedtii*)	53.9	24.1	19.7	-	-	0.9	3.4–3.9*	Lopez et al. (2020)
	—	52.0	24.0	14.6	4.6	-	2.6	-	Mol and Turan (2008)
	—	-	-	-	-	-	3.4–3.8	-	Wirth et al. (2002)
	—	50.1	25.9	15.4	-	-	1.9	-	Shuang et al. (2019)
	—	-	-	-	-	-	6.2	-	Ma et al. (2020)
	Sevruga (*Acipenser stellatus*)	-	-	-	-	-	2.0–3.2	2.1–2.8*	Wirth et al. (2002)
	Siberian (*Acipenser baerii*)	59.5	23.8	14.9	-	-	1.4	2.4–4.9*	Lopez et al. (2020)
	—	51.8	24.0	14.2	-	-	1.14	-	Gong et al. (2003)
	—	-	-	-	-	-	2.2–4.5	-	Wirth et al. (2002)
	—	64.0	19.8	10.5	-	-	3.3	2.5	Kowalska-Góralska et al. (2020)
	—	52.1	25.4	14.9	-	-	4.2	-	Shuang et al. (2019)
	—	-	-	-	-	-	6.3	-	Ma et al. (2020)
	Sterlet (*Acipenser ruthenus*)	51.3	25.4	13.2	-		1.7	1.9–2.5*	Park et al. (2015)

Tilapia	*Oreochromis niloticus*	52	-	-	-	-		2.4	Gunasekera et al. (1996)
Trout	*Oncorhynchus masou*	57.3	28.0	12.6	-	-	-	-	Suzuki and Suyama (1983)
	Oncorhynchus mykiss (Salmo gairdneri)	63	22.5	5.0	-	-	1.5	-	Baki et al. (2021)
	—	59.2	23.8	12.5	-	-	2.5	-	Vasconi et al. (2020)
	—	-	-	9.2	-	0.29	2.4	-	Kaitaranta and Ackman (1981); Kaitaranta and Linko (1984)
	—	63.7	27.4	7.6	-	-	-	-	Vúorela et al. (1979)
	—	64	24.3	9.1	-	0.07	-	-	İnanlı et al. (2019)
	—	55.6	30	11.8	-	-	-	-	Suzuki and Suyama (1983)
	—	-	-	6.5–8.8	-	0.20	5.9	-	Bledsoe et al. (2003); Bledsoe and Basco (2006)
	—	54.8	30.6	9.0	-	0.30	1.5	-	Kalogeropoulos et al. (2012)
	—	64.0	23.8	4.6		-	3.3	4.5	Kowalska-Góralska et al. (2020)
	Salmo trutta (sea trout)	63.5	24.9	4.3	-	-	-	5.3	Kowalska-Góralska et al. (2020)
Tuna	*Thunnus tonggol*	72.2	18.4	5.7	1.55	0.12	5.5	-	Intarasirisawat et al. (2011)
Turbot	*Scophthalmus maximus*	-	3.7	-	-	-	-	1	Rønnestad et al. (1993)
Vendace	*Coregonus albula*	-	-	9.8	-	0.14	3.4	-	Kaitaranta and Ackman (1981); Kaitaranta and Linko (1984)
	—	69.3	18.7	9.9	-	-	-	-	Vúorela et al. (1979)
	—	-	-	7–10	-	-	2.8	0.9–1.4	Bledsoe et al. (2003); Bledsoe and Basco (2006)

(continued)

Fish									
Species	Scientific name	Water	Protein	Lipid	Carb	Chol	ω-3/ω-6	*d*	References
White seabream	*Diplodus sargus*	92.7	-	1.7	-	0.21	8.7	-	Cejas et al. (2004)
	—	-	-	-	-	-	-	0.8–0.9	Sanchez-Velasco and Norbis (1997)
Whiting	*Merlangius merlangus*	85	-	2.8	-	0.33	14.6	1.0–	Tocher and Sargent (1984)
	—	76–84	7–15	0.4–9.7	1.1–2.8	-	-	1.1	Tavakoli et al. (2021)
Wolffish	*Anarhichas lupus*	-	-	-	-	-	-	5.0–6.0	Ehrenbaum and von Fischen (1905-1909)

Crustaceans									
Species	Scientific name	Water	Protein	Lipid	Carb	Chol	ω-3/ω-6	*d*	References
Crab	*Cancer pagurus*	57	25	3.3	-	0.17	4.2	-	Barrento et al. (2010)
	Paratelphusa hydrodromous	-	-	-	-	-	-	0.8	Harrison (1990)
	Portunus trituberculatus	55.4	30.2	13.0	-	0.50	-	-	Bledsoe et al. (2003)
Lobster	*Homarus americanus*	68.5	22.1	4.4	-	0.11	5.4	-	Barrento et al. (2009)
	Homarus gammarus	58.9	24.1	3.8	-	0.11	4.1	-	Barrento et al. (2009)
Norway lobster	*Nephrops norvegicus*	50	22–30	6–11	1–2	-	-		Tuck et al. (1997)
	—	62.9	-	-	-	-	5.6	1.6	Rosa et al. (2003)
	—	-	-	-	-	-	-	1.4	Mori et al. (2001)

Shrimp and prawn	*Cryphiops caementarius*	75	12.5	7.1	-	-	-	-	Moreno-Reyes et al. (2015)
	Chorismus antarcticus	72	-	5.3	-	0.13	7.0	1.8	Graeve and Wehrtmann (2003)
	Nematocarcinus lanceopes	67	-	4.7	-	0.11	5.2	1.1	Graeve and Wehrtmann (2003)
	Netocragon antarcticus	70	-	5.4	-	0.22	4.1	1.3	Graeve and Wehrtmann (2003)
	Alpheus saxidomus	76	-	4.8	-	-	3.3	1.0	Wehrtmann and Graeve (1998)
	Palaemonetes schmitti	74	-	4.1	-	-	7.4	0.6	Wehrtmann and Graeve (1998)
	Palaemon elegans	60	-	6.1	-	0.8	5.2	0.8	Morais et al. (2002)
	Palaemon serratus	65	-	5.2	-	1.0	1.3	0.9	Morais et al. (2002)
	Plesionika martia martia	73	-	3.6	-	0.8	4.6	0.7	Morais et al. (2002)

(continued)

Mussels									
Species	Scientific name	Water	Protein	Lipid	Carb	Chol	ω-3/ω-6	*d*	References
Blue mussel	*Mytilus edulis*	95–98	-	-	-	-	-	-	Harbach and Palm (2018)
	—	78–88	-	-	-	-	-	-	Desnica et al. (2011)
	—	(93)	3–4	1.5	-	-	-	-	Pieters et al. (1980)
Oyster	*Crassostrea gigas*	(80)	-	3	-	-	7–10	-	Soudant et al. (1999)
	—	80.4	-	-	-	-	-	-	Schmidt et al. (2020)
	—	(80)	10	1	-	-	-	0.04	Ren et al. (2003)
	Ostrea edulis	75.5	-	-	-	-	-	-	Schmidt et al. (2020)
	—	(75.5)	12.3	2.9	8.6	2.0	2.5	-	Frolov and Pankov (1992)
Scallop	*Argopecten ventricosus*	-	-	1–2	-	0.1–	9	-	Arjona et al. (2008)
	Flexopecten glaber	79	13	3	-	0.4	-	-	Berik and Cankiriligil (2013)
	Nodipecten subnodosus	-	-	1–7	-	-	-	-	Palacios et al. (2007)
	Pecten maximus	-	4	1	0.5	0.3	2–5	-	Soudant et al. (1996)

Echinoderms									
Species	Scientific name	Water	Protein	Lipid	Carb	Chol	ω-3/ω-6	*d*	References
Sea cucumber	*Stichopus japonicus*	-	-	0.5–1.2	-	-	1.5	-	Kasai (2003)
Sea urchin	Unspecified	-	-	3.1	-	0.50	-	-	Krzynowek and Murphy (1987)
	Diadema sp.	(74)	15.1	7.0	3.9	-	-	0.9	McAlister and Moran (2012)
	Echinometra sp.	(74)	16.9	7.5	1.9	-	-	0.8	McAlister and Moran (2012)
	Eucidaris sp.	(74)	16.5	7.0	1.9	-	-	0.7	McAlister and Moran (2012)
	Hemicentrotus pulcherrimus	74.1	16.3	8.4	-	0.31	-	-	Bledsoe et al. (2003)
	Paracentrotus lividus	73	15.1	3.8	-	-	6.7	-	Cruz-García et al. (2000)
	—	70.5	16.5	7.8	-	0.30	0.4	-	Kalogeropoulos et al. (2012)
	—	79–82	10–12	2.3–4.2	2.0–2.5	-	-	-	Dincer and Cakli (2007)
	—	67–70	18–20	3.8–4.4	-	-	-	-	Camacho et al. (2023)
Starfish	*Echinaster* spp.	(75)	11–14	9–14	0.2–0.5	-	-	-	Scheibling and Lawrence (1982)

Cephalopods									
Species	Scientific name	Water	Protein	Lipid	Carb	Chol	ω-3/ω-6	*d*	References
Cuttlefish	*Sepia esculenta*	87	10	0.3	1.5	-	7.5	13	Lei et al. (2014)
	Sepia officinalis	-	-	3.4	-	0.19	-	-	Blanchier and Boucaud-Camou (1984)
	—	89	-	0.22	-	-	8.6	-	Sykes et al. (2009)

(continued)

Cephalopods									
Species	Scientific name	Water	Protein	Lipid	Carb	Chol	ω-3/ω-6	*d*	References
Octopus	*Eledone cirrhosa*	(77)	13.9	2.3	-	0.15	4.9	-	Rosa et al. (2004b)
	Eledone moschata	(77)	12.2	2.6	-	0.17	4.9	-	Rosa et al. (2004b)
	Octopus defilippi	(72)	16.9	-	-	0.17	4.9	-	Rosa et al. (2004a)
	Octopus ocellatus	77.1	17.3	-	-	0.19	-	-	Bledsoe et al. (2003)
	Octopus vulgaris	72	20	4.4	-	-	-	-	Valverde et al. (2013)
	—	(72)	-	3.1–3.6	-	0.23	4–5	-	Lourenço et al. (2014)
	—	(72)	17.7	-	-	0.12	2.7	-	Rosa et al. (2004a)
	—	65	23	2.2	-	0.4	3–15	1.8	Quintana et al. (2015)
Squid	*Doryteuthis bleekeri*	70.0	23.4	5.1	-	0.37	-	-	Bledsoe et al. (2003)
	Loligo vulgaris	-	-	-	-	-	28	-	Salman et al. (2007)
	—	-	-	-	-	-	-	2	Şen (2005)

Table 2. Proximate Composition of Processed Roe

Proximate composition of products of dried roe from gray mullet (*Mugil cephalus*). Contents of protein, lipid, and cholesterol are given in mg/100 g wet weight. ω-3/ω-6 denotes the ratio between total amount of ω-3 and ω-6 fatty acids. No data are available for carbohydrate content. "-" denotes that data is not available

Origin of product	Water	Protein	Lipid	Chol	ω-3/ω-6	References
Greece	45.1	33.6	17.7	0.28	4.55	Kalogeropoulos et al. (2008)
Greece	48.1	33.7	17.8	0.51	3.2	Kalogeropoulos et al. (2012)
Florida	30.5	35.5	25.7	-	3.0	Kalogeropoulos et al. (2008)
Japan/Nagasaki	23.4	38.7	33.1	-	-	Kalogeropoulos et al. (2008)
Taiwan	25.2	44.2	26.2	-	-	Kalogeropoulos et al. (2008)
Italy	25.2	42.8	29.7	-	-	Kalogeropoulos et al. (2008)
Italy	-	-	-	-	5.0	Scano et al. (2008)

Table 3. Free Amino Acid Profiles

Free amino acid profiles are given in units of mg/100 g wet weight for selected types of marine roe. Results are shown for a fatty fish (salmon, *Oncorhynchus keta*), a lean fish (cod, *Gadus morhua*), a crustacean (king crab, *Paralithodes camtschaticus*), an echinoderm (*Paracentrotus lividus*), and a cephalopod (squid, *Loligo forbesii*). For comparison, the profile is also given for chicken egg yolk. "-" denotes that data is not available. *Assumed moisture of chicken egg yolk: 53%

Free amino acid (FAA)	Fatty fish Salmon (*Oncorhynchus keta*) ([1]Suzuki & Suyama, 1983; [2]Raza et al. unpubl.)	Lean fish Cod (*Gadus morhua*) (Raza et al. unpubl.)	Crustacean King crab (*Paralithodes camtschaticus*) (Raza et al. unpubl.)	Echinoderm Sea urchin (*Paracentrotus lividus*) (Camacho et al., 2023)	Cephalopod Squid (*Loligo forbesii*) (Raza et al. unpubl.)	Chicken egg yolk* (Nimalaratne et al., 2011)
Asp	12.9[1], 50.4[2]	10.4	8.1	120	120.4	24
Glu	17.6[1], 28.8[2]	53.6	23.9	210	157.2	55
Ser	8.2[1], 15.2[2]	27.7	25.4	70	63.9	28
His	0.9[1], 5.0[2]	15.8	25.6	90	37.2	7
Gly	1.7[1], 4.8[2]	26.3	102.5	1,260	186.8	12
Thr	4.7[1], 8.2[2]	23.4	33.1	65	69.5	27
Arg	2.9[1], 1.7[2]	23.8	133.9	570	108.3	74
Ala	5.0[1], 13.1[2]	50.9	41.5	140	130.4	30
Tyr	3.0[1], 4.2[2]	16.1	26.9	160	82.6	32
Val	6.1[1], 26.9[2]	22.4	21.6	70	87.8	31
Met	2.2[1], 14.6[2]	23.6	25.1	25	68.6	6
Trp	+[1], 4.8[2]	8.6	34.9	25	33.5	2
Phe	3.6[1], 3.2[2]	14.7	21.9	30	128.7	16
Ile	4.9[1], 4.4[2]	10.2	18.1	30	87.9	18
Leu	5.3[1], 7.2[2]	36.0	41.4	50	216.5	35
Lys	3.5[1], 13.0[2]	35.4	93.1	320	81.3	41
Pro	1.1[1], -[2]	-	-	30	-	-
Cys	0.6[1], -[2]	-	-	100	-	-
Tau	36.1[1], -[2]	-	-	40	-	-
Asn	13.1[1], -[2]	-	-	15	-	14
Gln	12.7[1], -[2]	-	-	130	-	16
Sum FAA	146[1], 205.6[2]	398.9	677	3550	1660.7	470

Table 4. Umami Synergy in Selected Roe and Gonads

Content (mg/100 g wet weight) of free glutamate (Glu) and certain free nucleotides (IMP, GMP, AMP), all of which are important for umami flavor, for a number of selected species. "-" indicates that no data are available. The equivalent umami intensity is given by a special number, EUC (mg/100 g).

Classical empirical studies have shown that the synergy between free glutamate and free nucleotides (N) with the concentrations u and v_N, respectively, can be expressed by the formula $\text{EUC} = u + u \times \Sigma_N \gamma(N) v_N$, where the sum Σ_N runs over all the different types of nucleotides present. $\gamma(N)$ is a set of empirically determined constants that characterize the strength of the synergy for the given nucleotides. The EUC is called the equivalent umami concentration and can be interpreted as the equivalent concentration of pure glutamate required to give the same umami intensity as the given mixture of glutamate and nucleotides, e.g., adenosine 5′-monophosphate (AMP), guanosine 5′-monophosphate (GMP), and inosine 5′-monophosphate (IMP). The strength factors have been determined experimentally to be $\gamma(\text{AMP}) = 0.22$, $\gamma(\text{GMP}) = 2.8$, and $\gamma(\text{IMP}) = 1.2$, and their stated values assume that all concentrations in the equation above are expressed in units of mg/100 g.

The range of species and subspecies in the table reflects that a significant part of the existing analyses has been carried out on species from the waters around Japan. The quoted values are subject to some uncertainty, among other things because they may depend on the degree of maturation of the eggs. For example, for crustaceans, it has been found that the content of free glutamate increases sharply as the embryo in the egg develops. Not all umami components have been measured in all cases, and the quoted values of EUC may therefore be underestimated. Abbreviations: Glu, glutamate; AMP, adenylate (adenosine 5′-monophosphate); GMP, guanylate (guanosine 5′-monophosphate); IMP, inosinate (inosine 5′-monophosphate)

Species	Scientific name	Glu	IMP	GMP	AMP	EUC	References
Alaska pollock	*Gadus chalcogrammus*	71	17	7	3	3000	Chiou et al. (1989)
Blue mussel	*Mytilus edulis*	75	8	2.5	25	1700	Ma et al. (2020)
Mullet	*Mugil cephalus*	38	5	3	<0.4	600	Chiou and Konosu (1988)
Norway lobster	*Nephrops norvegicus*	60	-	-	-	[60]	Rosa et al. (2003)
Salmon	*Oncorhynchus keta*	3–19	6–18	4	12–18	500	Hayashi et al. (1990)
Sea urchin	*Hemicentrotus pulcherrimus*	318	7	6	32	5300	Konosu (1973)
	Paracentrotus lividus	210	70	70	19	60,000	Camacho et al. (2023)
Sturgeon	*Unspecified*	82	4.4	-	-	[100]	UIC (2025)
Chicken egg yolk	*Gallus domesticus*	55	-	-	-	[55]	Nimalaratne et al. (2011)

Table 5. Development of Umami Compounds in *Bottarga*

Contents of free glutamate (Glu; mg/100 g) and free nucleotides (IMP, GMP, AMP; mg/100 g) during the production of *bottarga* from mullet roe. The last step from semi-dried to final product takes place in sunlight. EUC denotes equivalent umami concentration and thus includes the synergy between glutamate and the free nucleotides. Data from Chiou and Konosu (1988). "-" denotes that data is not available

	Fresh roe	Salted	Semi-dried	Final product
Glu	38	80	84	158
IMP	5	4	5	12
GMP	3	2	4	7
AMP	<0.4	-	-	1
EUC	600	900	1,500	5,500

Table 6. Fatty Acid Profiles

Fatty acid profiles for selected types of marine roe. Data correspond to percentage (g/100 g) of total fatty acid content. Results are shown for a fatty fish (salmon, *O. keta*), a lean fish (cod, *G. morhua*), a crustacean (Norway lobster, *Nephrops norvegicus*), an echinoderm (*P. lividus*), and a cephalopod (cuttlefish, *Sepia officinalis*). For comparison, the profile is also given for chicken egg yolk. The nomenclature used, C X:Y, is the standard one with X referring to carbon chain length (number of carbon atoms) and Y the number of double bonds. Further specification is given as ω-m, m = 1,3,4,5,6,7,8,9,11, for the unsaturated fatty acids. AA, arachidonic acid (ω-6); ALA, α-linoleic acid (ω-3); DHA, docosahexaenoic acid (ω-3); EPA, eicosapentaenoic acid (ω-3); LA, linoleic acid (ω-6). ω-3/ω-6 denotes the ratio between total amount of ω-3 and ω-6 fatty acids. "-" denotes that data is not available

Fatty acid	Fatty fish Salmon (Vasconi et al., 2020)	Lean fish Cod (Vasconi et al., 2020)	Crustacean Norway lobster (Rosa et al., 2003)	Echinoderm Sea urchin (Cruz-García et al., 2000)	Cephalopod Cuttlefish (Sykes et al., 2009)	Chicken egg (Frida, 2025)
C 12:0	-	-	0.15	-	-	0.0
C 14:0	3.44	1.47	1.8	11	2.8	0.31
C 15:0	0.54	0.30	0.69	-	-	0.0
C 15:1	-	-	-	-	-	-
C 16:0	10.63	17.59	18	20	25	24.2
C 16:1 ω-5/7/9	4.83	4.89	8.3	3.2	2.3	2.28

(continued)

Fatty acid	Fatty fish Salmon (Vasconi et al., 2020)	Lean fish Cod (Vasconi et al., 2020)	Crustacean Norway lobster (Rosa et al., 2003)	Echinoderm Sea urchin (Cruz-García et al., 2000)	Cephalopod Cuttlefish (Sykes et al., 2009)	Chicken egg (Frida, 2025)
C 16:2 ω-4	0.14	-	-	-	-	-
C 16:3 ω-3	-	-	-	-	0.45	-
C 16:3 ω-4	0.85	0.08	-	-	-	-
C 16:4 ω-3	-	-	0.074	-	-	-
C 17:0	0.51	0.25	0.63	0.18	1.7	-
C 17:1 ω-8	--	-	1.0	-	-	-
C 18:0	4.76	2.83	3.4	3.3	9.2	9.0
C 18:1 ω-9	18.25	12.50	27	2.0	3.1	42.5
C 18:1 ω-7	2.85	4.25	5.7	3.0	2.2	1.89
C 18:1 ω-5	-	-	-	-	0.51	-
C 18:2 ω-7	-	-	-	-	-	-
C 18:2 ω-6 (LA)	1.32	0.53	1.1	0.95	0.81	13.8
C 18:3 ω-3 (ALA)	1.22	-	0.49	1.9	0.61	0.84
C 18:3 ω-6	-	-	0.074	0.28	-	-
C 18:4 ω-3	1.07	0.46	0.35	3.6	-	0.0
C 18:4 ω-1	0.34	-	-	-	-	-
C 19:0	-	-	0.29	-	-	-
C 20:0	-	-	0.27	0.98	0.85	0.0
C 20:1 ω-9/11	1.28	4.48	-	8.5	2.7	0.46
C 20:1 ω-7	0.32	0.34	0.70	-	-	-
C 20:2 ω-6	0.32	-	0.52	1.5	0.15	0.033
C 20:3 ω-6	0.12	-	0.29	0.43	-	0.025
C 20:3 ω-3	-	-	0.25	15	0.23	0.96
C 20:4 ω-6 (AA)	1.42	3.07	2.0	2.7	3.5	2.1

(continued)

Fatty acid	Fatty fish Salmon (Vasconi et al., 2020)	Lean fish Cod (Vasconi et al., 2020)	Crustacean Norway lobster (Rosa et al., 2003)	Echinoderm Sea urchin (Cruz-García et al., 2000)	Cephalopod Cuttlefish (Sykes et al., 2009)	Chicken egg (Frida, 2025)
C 20:4 ω-3	2.51	0.12	0.59	-	-	0.0
C 20:5 ω-3 (EPA)	15.41	17.80	7.7	16	15	0.013
C 21:5 ω-3	-	-	-	-	0.07	-
C 22:0	-	-	0.09	-	-	0.086
C 22:1 ω-11	-	-	0.63	-	0.0	0.0
C 22:1 ω-9	0.19	-	0.28	3.7	0.0	0.0
C 22:4 ω-6	-	-	0.14	-	0.11	2.05
C 22:5 ω-3	5.56	1.90	1.3	-	1.4	0.042
C 22:5 ω-6	-	-	0.38	-	0.62	-
C 22:6 ω-3 (DHA)	22.04	27.14	15	2.5	27	0.16
ω-3/ω-6	15.14	13.20	5.6	6.7	8.6	0.13

Table 7. Energy Content of Selected Fresh and Processed Roe

Energy content in both kJ/100 g and kcal/100 g of roe from a selection of marine species as well for a processed mullet roe product (*bottarga*). For comparison, data is also given for chicken egg yolk + white

Species	Scientific name	Energy (kJ/100 g)	Energy (kcal/100 g)	References
Capelin	*Mallotus villosus*	770	185	Frida (2025)
Cod	*Gadus morhua*	473	113	Krzynowek and Murphy (1987)
	—	502	120	Frida (2025)
Herring	*Clupea harengus*	448	107	Kalogeropoulos et al. (2012)
	—	335	80	Krzynowek and Murphy (1987)
Lumpfish	*Cyclopterus lumpus*	552	132	Kalogeropoulos et al. (2012)
	—	443	106	Frida (2025)

(continued)

Species	Scientific name	Energy (kJ/100 g)	Energy (kcal/100 g)	References
Mullet (*bottarga*)	*Mugil cephalus*	1300–1900	310–450	Mol and Turan (2008); Kalogeropoulos et al. (2008)
	—	1264	302	Kalogeropoulos et al. (2012)
Rainbow trout	*Oncorhynchus mykiss*	782	187	Machado et al. (2016)
Salmon	*Salmo salar*	1273	304	Mol and Turan (2008)
	—	1230	294	Kalogeropoulos et al. (2012)
Sturgeon	Different species	1220–1330	300–320	Mol and Turan (2008)
Sea urchin	*Paracentrotus lividus*	678	162	Kalogeropoulos et al. (2012)
Whiting	*Merlangius merlangus*	64–154	20–40	Tavakoli et al. (2021)
Garden snail	*Cornu aspersum*	185	44	Maćkowiak-Dryka et al. (2020a, b)
Chicken egg (yolk + white)	*Gallus domesticus*	1340	324	Frida (2025)

Table 8. Aroma Compounds in Roe

Typical aroma compounds found in fish roe, their characteristic odor, and their chemical origin. Data from Vilgis (2020)

Aroma compound	Characteristic odor	Chemical origin
(E,Z)-2,4-Decadienal	Green, roasted, oily, fishy	Fatty acids
(E,Z)-2,6-Nonadienal	Green, cucumber-like, melon-like	Fatty acids
(E,E)-2,4-Decadienal	Green, oily, chicken skin, soup-like	Fatty acids
2-Nonenal	Green, waxy, fatty, melon-like	Fatty acids
Nonanal	Green, floral, waxy, citrus-like	Fatty acids
(Z)-4-Heptanal	Green, oily, milk-like, creamy, cheesy	Fatty acids
Methional	Sulfuric, yeasty, cheesy, aged, boiled	Amino acids
Octanal	Green, waxy, citrus-like, fishy	Fatty acids
Hexanal	Green, oily, leafy	Fatty acids
1-Octen-3-ol	Green, mushroom-like, musty	Fatty acids
2-Pentylfurane	Caramel-like, fruity, green, earthy	Carbohydrates
Phenylacetaldehyde	Aromatic, honey-like, spicy, sweet, floral	Amino acids

Table 9. Off-Flavors in Roe

List of unpleasant aromas in roe (off-flavors), their chemical origin, and in which types of roe they may occur. Data from Sicuro (2019)

Aroma/smell	Chemical origin	Species
Fishy	Aldehydes	White sturgeon
Earthy, musty	Geosmin, isoborneol	Rainbow trout
Earthy, woody, fishy, rancid, rotten, petroleum, musty	Geosmin, isoborneol	Salmon, carp, catfish, tilapia
Cooked-potato-like	Strecker aldehydes	Cod
Muddy	Geosmin, isoborneol	Rainbow trout, catfish, tilapia
Muddy, earthy	Geosmin, isoborneol	White sturgeon, sweet water bream

Table 10. Additives in Some Roe Products

List of possible additives in commercial roe products, such as stabilizers, preservatives, and coloring and flavoring agents. All allowed additives must be declared by the E-number under EU regulations. Data from Vasconi et al. (2020)

Additive	E-number and chemical name
Stabilizers and thickeners	E 412 (guar gum); E 413 (tragacanth); E 415 (xanthan gum); E 422 (glycerol)
Preservatives	E 200 (sorbic acid); E 202 (potassium sorbate); E 211 (sodium benzoate)
Antioxidants	E 330 (citric acid); E 331 (sodium citrate)
Coloring agents	E 102 (tartrazine, gul); E 120 (carmines, red); E 133 (brilliant blue); E 141 (chlorophyll, green-blue-black); E 150d (ammonia caramel, brown); E 151 (brilliant black); E 160a (carotenes, yellow); E 160c (paprika extract, red); E 163 (anthocyanin, red violet); safflower (yellow)
Tastants	E 621 (sodium glutamate, umami); E 622 (potassium glutamate, umami); soy sauce (salt, umami); sugar (sweet)

Table 11. Characteristics and Typical Uses of Selected Types of Roe

Egg size, color, and typical uses of roe from selected fish, crustaceans, and echinoderms. Data from Table 1, McGee (2004), and Joachim and Schloss (2008)

Species	Egg diameter *d* (mm)	Color	Uses
Alaska pollock	1.3–1.5	Light red, gray	*Mentaiko*
Capelin	1.0–1.2	Orange, yellow	*Masago*-kaviar, sushi
Carp	0.8–1.6	Light red	*Taramasalata*
Flatfish	1–1.5	Yellow, gray	Fried and boiled roe sacs
Cod	1.3–1.4	Light red, gray	Boiled or smoked roe sac, *taramasalata*, Kalles Kaviar
Flying fish	0.5–0.8	Golden, orange	*Tobiko*-kaviar, sushi
Herring	1.3–1.5	White yellow, golden	*Kazunoko, kazunoko konbu*
Lobster	0.4	Black or red (boiled)	In soups
Lumpfish	2–5	Light red	Kaviar
Mullet	0.6	Yellow	*Bottarga*
Salmon	4–7	Orange, red	*Ikura, sujiko*, kaviar, sushi
Sardine	1.1–1.3	Golden	Salted, smoked, in *dashi*
Sea urchin	0.7–0.9	Yellow, brownish	Sushi, in soups
Sturgeon	1–3.5	Black, gray, golden	Caviar
Trout	3.5–4.5	Red orange	Kaviar
Tuna	0.9–1.1	Yellow	*Bottarga*
Vendace	0.9–1.4	Light red, golden	*Lögrom*
Whitefish	0.9–1.4	Light red, golden	*Sikrom*

The Illustrations of the Book

Unless otherwise noted below, all photographs in the book are by Jonas Drotner Mouritsen. SS indicates images from Shutterstock.

Figure 1.4: Gadira, Productos de Almadraba; Fig. 2.3: SS Spalnic; Fig. 3.1: Unidentified source; Fig. 3.2: SS StLynx; Fig. 3.4: SS FreeProd33; Fig. 3.5 Groupe KAVIAR; Fig. 3.6: Anna Claire Heraud; Fig. 3.7: Groupe KAVIAR; Fig. 3.8: Anna Claire Heraud; Fig. 3.10: Anna Claire Heraud; Fig. 3.12: Ole G. Mouritsen; Fig. 5.2: Ole G. Mouritsen; Fig. 5.3: Ole G. Mouritsen; Fig. 5.6: Ole G. Mouritsen; Fig. 5.9: Ole G. Mouritsen; Fig. 6.1: Ole G. Mouritsen; Fig. 6.2: SS vandycan; Fig. 6.3: SS Richard Griffin; Fig. 6.4: Ole G. Mouritsen; Fig. 6.5: SS Gitanna; Fig. 6.6: Ole G. Mouritsen; Fig. 7.1: Ole G. Mouritsen; Fig. 7.3: (left) SS Philip Garner, (right) SS BlueTrend; Fig. 7.4: Ole G. Mouritsen; Fig. 7.5 SS 183614594 Alex Staroseltsev; Fig. 7.6: SS Aksornsawan Seebun; Fig. 7.7: Claes Bech-Poulsen; Fig. 8.1: Siriusplot. Creative Commons Attribution 2.0 Generic license.https://commons.wikimedia.org/wiki/File:Kuchiko_Sado_Japan1.jpg; Fig. 8.2: Naotake Murayama. Creative Commons Attribution 2.0 Generic license. https://www.flickr.com/photos/12832970@N00/2678940158; Fig. 8.3: SS TKHS; Fig. 9.2: SS Praisaeng; Fig. 9.3: Cavi-art; Fig. 9.4: SS margouillat photo; Fig. 10.1: SS karins; Fig. 10.2: Gadira, Productos de Almadraba; Fig. 11.2: Ole G. Mouritsen; Fig. 11.5: Ole G. Mouritsen.

O. G. Mouritsen, K. Styrbæk, *Roe and Roe Gastronomy*,
https://doi.org/10.1007/978-3-032-13142-3

Glossary of Technical and Scientific Terms

AA See arachidonic acid.

Adenosine triphosphate (Adenosine-5′-triphosphate, ATP) nucleotide, which is the biochemical source of energy production in living cells. Can be broken down, among other things, into the 5′-ribonucleotides, inosinate, adenylate, and guanylate, which are associated with synergy in the taste of umami.

Adenylate (AMP, adenyl-5′-monophosphate) nucleotide, which in free form together with glutamate provides synergy in the taste of umami, e.g., in certain types of roe.

Adhesion Sticking together.

Albumin Protein in egg white.

Aldehyde Organic chemical compound, which is often found in flavorings, which in roe, e.g., can give a fishy or earthy smell and taste.

Almas (Alma; Persian, diamond) rare, whitish-yellowish and dull caviar from albino sturgeons, which have no pigments.

Alginate Polysaccharide, found in brown seaweeds. Alginates form gels in the presence of Ca^{++} (or other divalent ions such as Mg^{++} and Ba^{++}) and can be used for spherification, e.g., in the production of kaviar substitutes.

Alginic acid Alginate in acid form, which is insoluble in water.

Alum Water-soluble potassium-aluminum salt ($KAl(SO_4)_2$), which has a sweet–sour, metallic and slightly astringent taste. Used, among other things, for marinating foods, e.g., to preserve the freshness of roe.

Amino acid Small, organic molecule with an amino group ($–NH_2$). Amino acids are the elementary building blocks of proteins. Examples are glycine, glutamic acid, alanine, proline, and arginine. Of the 20 natural amino acids, nine are so-called essential amino acids, which our body cannot produce itself and which we have to obtain from our diet (valine, leucine, lysine, histidine, isoleucine, methionine, phenylalanine, threonine, and tryptophan). In foods, amino acids are mainly found bound in proteins but to some extent also as free amino acids that

O. G. Mouritsen, K. Styrbæk, *Roe and Roe Gastronomy*,
https://doi.org/10.1007/978-3-032-13142-3

can affect taste, e.g., glutamic acid, which is the basis of umami taste. Histidine has a bitter taste. Many of the free amino acids in roe are of the type associated with the taste of seafood (arginine, glutamic acid, alanine, and glycine).

AMP See adenylate.

Anadromous Term for fish that live in river deltas and migrate up rivers to spawn. Salmon and most sturgeons are anadromous fish.

Anchoviar Product name for Spanish caviar substitute made from anchovies.

Antioxidant Substance that can prevent the oxidation of other substances. Ascorbic acid (vitamin C), citric acid, and carotenoids, e.g., carotene, can act as antioxidants. Antioxidants added to roe can inhibit oxidation and thus rancidity of the unsaturated fatty acids in the roe.

Aquaculture Cultivation of marine organisms in farms in the sea or in ponds on land.

Arachidonic acid (AA) super-unsaturated, long-chain fatty acid with 20 carbon atoms and four double bonds. Belongs to the ω-6 family.

Armored eel A group of fish that, together with bichirs, belong to an ancient order (Polypteriformes) of ray-finned fish that was formed at the same time as the order of sturgeons (Acipenseriformes).

Aroma Odorant that can be detected by receptors in the mucous membranes of the nose.

Aspartic acid Amino acid, whose free salts (aspartates) can contribute to the umami taste like glutamate, although much less intensively.

Astacene Red-orange dye, found in certain shellfish and roe.

Astaxanthin Red-colored carotenoid, e.g., found in shellfish and roe. As long as astaxanthin is bound in the protein complex crustacyanin, it does not show its red color.

Avruga Product name for Spanish caviar substitute made from smoked herring fillet.

Barako By-product of *sujiko* production consisting of the loose salmon eggs that may have broken free from the roe mass during the process. *Barako* is used in sushi and as a filling in rice balls (*onigiri*).

Berries Popular term for the eggs that crustaceans hold between their small swimlets while the eggs mature.

Bichir A group of fish that, together with the armored eels, belong to an ancient order (Polypteriformes) of ray-finned fish that was formed at the same time as the order of sturgeons (Acipenseriformes).

Biogenic amines Toxic substances, e.g., histamine, that can be formed in marine products such as roe by the action of enzymes and microbial activity.

Borax Inorganic salt (sodium tetraborate, $Na_2[B_4O_5(OH)_4]\cdot 8H_2O$) of boric acid, declared as the additive E 285 when used as a preservative in, for example, roe products.

Boric acid Is used as a preservative in the form of the salt borax.

Bottarga (*Botargo*) dried fish roe from, for example, tuna, cod, or mullet. A great delicacy in Spain (*botargo*) and in Italy (*bottarga*). Has lots of umami and is eaten as tapas or with a pasta dish.

Brilliant black (Brilliant Black BN) synthetic tar dye of the azo type, E 151, which is used, among other things, to color roe products black.

Calciferol Vitamin D_3.

Canthaxanthin Carotenoid with a greenish color and which is a pigment in certain types of roe.

Carbohydrates (Saccharides or sugars) are a large group of organic compounds that mainly contain oxygen, hydrogen, and carbon. The simple saccharides, monosaccharides and disaccharides, are sweet and include the common sugars such as monosaccharides glucose, fructose, and galactose, as well as disaccharides sucrose, lactose, and maltose. More complex carbohydrates (polysaccharides), which are often tasteless, can act as both a structure builder and an energy store.

Carotenoid Group of red-orange color pigments in plants and animals, e.g., astaxanthin in fish and shellfish and carotene in carrots. In roe there is a wide range of carotenoids, e.g., astaxanthin, canthaxanthin, zeaxanthin, diatoxanthin, doradexanthin, idoxanthin, tunaxanthin, and lutein, as well as substances derived from astaxanthin. Carotenoids are fat-soluble.

Carrageenan Polysaccharide found in red seaweed species, which is suitable as a thickener.

Caviar Protected name for roe products from sturgeon.

Cavi-art Product name for kaviar substitute made from alginate from brown seaweed species.

Cephalopods Class (Neocoloidea) of higher mollusks that include eight-armed (e.g., octopus) and ten-armed squid and cuttlefish.

Chawan mushi Traditional Japanese egg cream made from *dashi*, in which sea urchin roe (*uni*) can been deposited.

Chromosome Hereditary material; part of the genome.

Chorion Wall of connective tissue (zona radiata) that forms a rigid network on the inside of the outermost membrane of an egg. The wall gives the egg shape and elasticity. The chorion is pierced by some pores and also has a special, narrow channel (micropyle) through which the sperm cells can penetrate when fertilizing the egg.

CITES (Convention on International Trade in Endangered Species of Wild Fauna and Flora) international species register, which is intended to ensure the legality of international trade in, for example, caviar from sturgeon. All trade in caviar during import and export must be clearly marked in accordance with this system. This applies to all types of caviar, including pressed and pasteurized caviar. The CITES rules presently only apply to caviar and no other types of fish roe.

Coagulation Process by which something clumps together (coagulates), e.g., proteins.

Cholesterol Special type of fat.

Codex Alimentarius International system for food concerning production, labeling, and safety.

Collagen Protein network that forms connective tissue and thus provides structure and coherence in cellular structures and tissues.

Collagenase Enzyme that can dissolve collagen fibers, e.g., in roe, so that the individual eggs are separated from each other.

Connective tissue (Collagen) network of collagen fibers, which are proteins.

Coral Gastronomic term for gonads and roe on, e.g., mussels and crustaceans.

Cross-linking Formation of chemical bonds crisscrossing between long-chain polymers, e.g., proteins or carbohydrates.

Crustacean Large suborder (Crustacea) of ten-legged arthropods with several families, where we find many of the so-called shellfish, e.g., lobsters, Norway lobsters, crabs, crayfish, shrimp, and prawn.

Crustacyanin Blue-green or reddish-brown protein complex, such as in the shells of crustaceans or in roe, where it is bound to the orange-red substance astaxanthin. When broken down during heating, the crustacyanin denatures, and the red color of the astaxanthin emerges.

Dashi Japanese term for a stock (jap. *Dashi*, boiled extract) made as an extract of seaweed (*konbu*) and skipjack fish flakes (*katsuobushi*). *Dashi* is the epitome of umami and umami synergy.

Demersal Term for fish that live near the seabed and lay their eggs on the seabed in contrast to pelagic (free-swimming) fish.

DHA See docosahexaenoic acid.

DNA (Deoxyribonucleic acid) polynucleotide consisting of a long chain of nucleotides (see nucleotide) formed from the sugar group deoxyribose and the bases adenine, guanine, thymine, and cytosine. DNA is the basis of the genetic information in the genome and thus the genetic material.

Docosahexaenoic acid (DHA) super-unsaturated, long-chain fatty acid with 22 carbon atoms and six double bonds. Belongs to the ω-3 family.

Donburi Japanese dish where various accessories, e.g., roe, are placed in a bowl of boiled rice.

Ebiko Japanese term for shrimp roe.

Echinoderms Order of invertebrates, of which there are approximately 6,000 different living species, which include, for example, the classes of sea urchins, starfish, and sea cucumbers. They are characterized by having an external armor in the form of a calcareous shell or a leathery thick skin.

Egg sac The membrane in which the roe are found in fish.

Eicosapentaenoic acid (EPA) super-unsaturated, long-chain fatty acid with 20 carbon atoms and five double bonds. Belongs to the ω-3 family.

Enzyme Special protein that acts as a catalyst in a chemical or biochemical reaction.

EPA See eicosapentaenoic acid.

Escargot pearls Eggs from land snails.

Fat Collective term for a large class of substances that cannot be dissolved in water. Fats can be solid, e.g., butter and wax, or liquid, e.g., olive oil and fish oil. The melting point of a fat has an impact on its taste, mouthfeel, and nutritional value. A typical fat consists of a long chain of carbon atoms. The chain can be saturated or unsaturated. An important type of natural fat is lipids, which are composed of fatty acids and various other substances, such as amino acids and saccharides. Unsaturated fats are found in large quantities in roe.

Fatty acid A fatty acid consists of a long chain of carbon atoms with a carboxylic acid group at one end. Neighboring carbon atoms in the chain are chemically linked together by either a single bond or a double bond. The more double bonds there are, the more unsaturated the fatty acid is. If there are only single bonds, the fatty acid is completely saturated. Monounsaturated fatty acids have one double bond, such as oleic acid from olive oil. Polyunsaturated fatty acids have more than one double bond, such as two double bonds in linoleic acid from soybeans or three double bonds in alpha-linolenic acid from flaxseed. Super-unsaturated fatty acids have more than four double bonds, e.g., six double bonds in DHA (docosahexaenoic acid), which is found in large quantities in roe. Essential fatty acids are fatty acids that our body cannot produce itself and that we therefore need to obtain from food. These are the polyunsaturated fatty acids linoleic acid and alpha-linolenic acid. These two types of unsaturated fatty acids are the progenitors of two important classes of fats, the ω-3 (e.g., EPA and DHA) and the ω-6 fats (e.g., AA).

Fecundity Term for the spawning ability and relative fertility of fish.

Filter feeders Marine animals that live on the seabed by filtering dissolved substances, particles, and microorganisms such as microalgae. Mussels are filter feeders.

Folate Vitamin B_6.

Follicle Structure of supporting cells around an egg.

Garum Brownish liquid formed by fermenting salted small fish, blood, and fish intestines of, e.g., anchovy, mackerel, and tuna. Closely related to Asian fish sauces.

Gashi Pakistani curry dish.

Genome The total genetic information (hereditary mass) in a given organism, i.e., all the genes.

Geosmin Organic chemical substance that, even in very small amounts, gives a fishy, musty, and earthy taste and smell. Geosmin is formed by various bacteria and can contribute to giving roe a less pleasant smell.

Glutamate Salt of the amino acid glutamic acid, e.g., monosodium glutamate (MSG). In not too acidic water, glutamic acid is found as glutamate ions. Glutamates are the main source of umami. Used as an additive in foods, MSG must be declared as E 621.

Glutamic acid Amino acid; the salts of glutamic acid are called glutamates, e.g., monosodium glutamate (MSG), which in ionic form elicits umami taste.

GMP See guanylate.

Gonads Reproductive organs of shellfish, crustaceans, mussels, and cephalopods.

Grain caviar Term for caviar of loose eggs in comparison to, for example, pressed caviar.

Green eggs Immature (not the color green) eggs of fish, e.g., sturgeon.

Guanylate (GMP, guanosine-5′-monophosphate) nucleotide, which together with glutamate provides synergy in the taste of umami, e.g., in certain types of roe.

Guar gum Gum extracted from the guar plant and used as a thickener.

Gunkan-sushi Sushi, where a small *nori* cuff is folded around rice, and the filling is placed on top (battleship sushi).

Hepatopancreas Fused organ in mollusks that has the functions of both the liver and the pancreas.

Histamine So-called biogenic amine, which is toxic and which is formed by microbial and enzymatic conversion of the amino acid histidine, e.g., by improper storage of roe products.

Histidine Amino acid, which can be converted to histamine by microbial activity.

Hormone Signaling substance that can stimulate cells to carry out certain biochemical processes.

Huevos de choco Nidamental glands of large cuttlefish of the species *S. officinalis*. The glands, which produce gelling agents to hold together and harden the eggs of the females, are a gastronomic specialty in Andalusia.

Hybrid Cross between two related species.

Hydrocarbon chain Organic compound with carbon and hydrogen in the form of a chain of carbon atoms, found in fats and fatty acids.

Hydrogel (Hydrocolloid) a partially solid material (a gel) in which large amounts of water are bound. Alginates from brown seaweed species are good for forming hydrogels for spherification of kaviar substitute products.

Ichigo-ni Japanese hotpot dish, e.g., with sea urchin roe.

Ikra Russian term for roe and kaviar from various types of fish, e.g., salmon.

Ikura Japanese term for roe and kaviar, typically from salmon. Derived from the Russian *ikra*.

IMP See inosinate.

Imperial Caviar Name for caviar from old osetra sturgeons, whose eggs retain a light color. This caviar is particularly sought after and therefore commands a higher price, although the color does not affect the taste.

Inosinate (IMP, inosine-5′-monophosphate) nucleotide, which in free form together with glutamate provides umami synergy, e.g., in certain types of roe.

Jastichnaja Low-quality caviar product, which is made from sturgeon eggs that are either far from maturation, are overripe, or that are poorly separated from the ovaries and therefore have more connective tissue.

Kaiseki Formal Japanese meal with many different dishes.

Karashi-mentaiko (*Mentaiko*) spicy variant of *tarako*, which is salted cod roe. The seasoning can be more or less strong, e.g., with garlic or Japanese chili (*togarashi*).

Karasumi Japanese term for salted and dried roe, e.g., *mentaiko*.

Kaviar Term for a product of loose eggs, e.g., from fish. Only kaviar from sturgeon may be called caviar.

Kazunoko Japanese term for whole, salted roe sacs from herring. *Kazunoko-konbu* is salted herring roe, which is spawned on the surface of *konbu* seaweed leaves.

Konbu (*Kombu*) large brown seaweed (*Saccharina japonica*), which is used, among other things, to make *dashi*. *Konbu* contains very large amounts of free glutamate and thus elicits umami.

Konoko See *kuchiko*.

Konowata Fermented and marinated sea cucumber intestines.

Kuchiko (*Konoko*) roe from sea cucumbers.

Lapti Romanian term for deep-fried carp milk.

Leucine Amino acid.

Lipid Fat, which consists of a water-soluble part and an oil-soluble part, which is usually a fatty acid. Biological membranes are made up of lipids, e.g., phospholipids and cholesterol.

Locust bean gum Substance extracted from the seeds of the locust bean tree and used as a thickener.

Lysine Amino acid.

Lysozyme Enzyme in egg white. Has antibiotic effect.

Lögrom Swedish term for kaviar from vendace.

Maillard reactions Class of chemical reactions typically associated with nonenzymatic browning by, e.g., frying, baking, and grilling. In these reactions, carbohydrates combine with amino acids from proteins, whereby after some steps a number of not particularly well-defined, brown, and aromatic substances are formed. These substances cause a wide range of taste and smell impressions, from floral and leafy to meaty and earthy.

Maki-sushi Sushi roll, in which there is a sheet of *nori* on the outside or inside.

Malossol Russian term meaning "lightly salted" and associated with the finest caviar, which only has 2.5–3% salt. Can also be used for other types of lightly salted roe.

Masago Japanese term for capelin roe.

Melanin Black pigment, found in dark roe and squid ink, among other things.

Membrane The boundary between a cell and its surroundings (cell wall). Used especially for the double layer of lipids (fats) that forms the middle part of the cell envelope.

Mentaiko Spicy variant of Japanese *tarako*, salted cod roe, e.g., *karashi-mentaiko.*

Merroir Term describing that the taste of marine products reflects their environment.

Metai Spicy roe; cheap American variant of *mentaiko*.

Methylisoborneol (2-MIB) strong and foul-smelling substance that, like geosmin, is produced by some microorganisms (soil bacteria), which are often found in land-based aquaculture installations and therefore in certain types of roe.

Micropyle Narrow channel in the collagen wall of an egg through which sperm can penetrate during fertilization.

Milt Sperm and sperm sacs from fish.

Mirin Sweet rice wine.

Mollusks Order (Mollusca) of animals that include cephalopods, mussels, and snails.

Moluga Product name for Spanish kaviar substitute and luxury product made from smoked fillet of North Sea herring and with a higher fat content than Avruga.

Monosodium glutamate See glutamate.

MSC (Marine Stewardship Council) international nonprofit organization that issues the MSC-Certified Sustainable Fisheries certificate.

MSG See glutamate.

Mouthfeel See texture.

Mussels Belong to the order of mollusks (Mollusca) and constitute a class (Bivalvia) of about 20,000 different species, most of which live in salt water. Examples are blue mussels, cockles, razor clams, scallops, and oysters.

Nigiri-sushi Hand-pressed sushi consisting of a rice ball with, e.g., a piece of fish or roe on top.

Nori Leaves of the red algae *Pyropia/Porphyra* spp., which are dried, pressed, and toasted in the form of paper-thin sheets that are used for *maki-sushi*, among other things.

Nucleic acid Chain of nucleotides (polynucleotide). Nucleic acids are the building blocks of DNA and genomes.

Nucleotide Chemical group that is part of nucleic acids. The umami substances inosinate, guanylate, and adenylate are nucleotides.

Omega-3 fats (ω-3 fats) see fats.

Omega-6 fats (ω-6 fats) see fats.

Onigiri Japanese term for balls of boiled rice with fillings of various kinds, e.g., roe.

Organoleptic Expression for the total sensory experiences of a food.

Osmosis Phenomenon that occurs across a wall, e.g., a cell membrane that is permeable to water but impermeable to other and larger molecules, e.g., salts, amino acids, or sugars. The imbalance that arises is compensated by some of the water penetrating to the side where the large molecules are.

Oursinado Traditional Provençal preparation with sea urchin roe, which is made by crushing the roe in a mortar and adding a little olive oil along the way, so that an emulsion is formed like a kind of mayonnaise.

Ovary Organ where the eggs of, e.g., fish are formed and stored.

Ovoglobulin Protein in egg white.

Ovoverdin Dark green, almost black protein complex in immature and mature crustacean roe. The complex also contains the red carotenoid astaxanthin, which is released when heated and colors the roe red.

Oxidation For example, unsaturated fats can be oxidized and thus go rancid.

Paddlefish A type of false sturgeon, of which there is only one living species, the American paddlefish, *Polyodon spathula*.

Panko Japanese breadcrumbs for breading.

Pasteurization Heat treatment technique to improve the shelf life of foods.

Payusnaya Russian term for pressed caviar, which contains up to 7% salt and is a kind of jam-like, spreadable paste with a stronger taste and aroma than caviar, which is mainly due to overripe and damaged eggs with oxidized fats.

PCR Biochemical technique for identifying DNA.

Pelagic Term for fish that live free-swimming and spawn their eggs in the water column (as opposed to demersal fish that move near the seabed).

Penaeid Term for a large family of crustaceans that includes tiger shrimps.

Phospholipid See lipid.

Pigment Dye, e.g., melanin in black roe and astaxanthin in red-orange roe.

Plankton Small animals, larvae, eggs, and algae that float freely in the oceans with the ocean currents.

Plastic Expression of the ability of materials (e.g., eggs) to yield when subjected to force, without the materials returning to their original shape when the force ceases.

Protein Polypeptide, i.e., a long chain of amino acids linked together by peptide bonds. A special class of proteins are enzymes, which cause chemical reactions to proceed under controlled conditions. Proteins lose their function (denature) and change their physical properties when heated or exposed to salt and acid or enzymatic action during fermentation. The breakdown of proteins produces smaller peptides and free amino acids, e.g., glutamic acid, which can provide flavor, e.g., umami taste.

Receptor Organ or protein molecule that has a special ability to recognize and bind a specific substance, e.g., an odor or taste molecule.

Riboflavin Vitamin B_2.

Roe Term for the collections of immature and mature eggs in the ovaries and possibly the mature eggs that have been released into the water, e.g., from fish, crustaceans, and shellfish.

Roe grains The individual eggs in a collection of roe, e.g., in an egg sac or released from the egg sac.

Spherification Process by which a spherical shell of a solid gel is formed around a liquid-filled volume. Spherification using alginate can be used to make kaviar substitutes, e.g., Cavi-art.

Shirako Japanese term for sperm and milt from cod.

Sikrom Kaviar from whitefish.

Shellfish Gastronomic term for mussels and crustaceans (arthropods).

Shovelnose sturgeon Small species of true sturgeon (*Scaphirhynchus platorynchus*).

Skrei Norwegian term for Atlantic cod.

Snails Large group of mollusks, from which the eggs of land snails are used as a kind of "white kaviar."

Sodium benzoate Preservative (E 211).

Sperm Male gametes, which in the form of milt in fish such as cod can be eaten as a kind of "white roe."

Sturddlefish Cross (hybrid) between Russian sturgeon and American paddlefish.

Sturgeon True sturgeon of the family Acipenseridae.

Sujiko Japanese term for whole roe sacs from salmon and trout.

Surimi Japanese term meaning minced meat. Used, for example, to imitate crab and shrimp meat using meat from lean fish.

Surströmming Swedish specialty made by fermenting the small Baltic herring *strömming*. The fermented herring are stored in cans, and you can often find whole roe sacs in the belly of the small herring. The taste is sour and has a lot of umami.

Swimlets Small limbs that sit on the underside of the tail of crustaceans. This is where the secreted eggs attach while they mature.

Tarako Japanese term for salted roe from especially cod fish. *Mentaiko* is a spicy version of *tarako*.

Tarama Greek term for fish roe.

Taramasalata Mixture of different types of roe, often cod roe (which can be smoked), and oil, with bread or mashed potatoes, lemon juice, and spices such as pepper and garlic added.

Tataki Japanese term for the treatment of fish or roe that are fried on the outside but raw on the inside.

Texture (Mouthfeel) the textural properties of food are the group of physical characteristics that are perceived primarily by the sense of touch and that are due to the structural properties of the food. Terms such as crunchy, firm, soft, and creamy are textural properties.

Tempura Japanese term for deep-fried fish, shellfish, or vegetables breaded in *panko*.

Thiamine Vitamin B_1.

Tobiko Japanese term for flying fish roe.

Transglutaminase Enzyme that catalyzes a special bond between a free amine group on a protein and the acyl group on the amino acid glutamic acid on another protein. The proteins are therefore firmly bound together in a way that they cannot be broken down by the enzymes (proteases) that normally break down proteins. Used, for example, to strengthen roe.

Trimethylamine Foul-smelling organic substance (tertiary amine), which, among other things, is formed by bacterial decomposition of trimethylamine oxide in dead fish but also in roe that are not fresh. Trimethylamine oxide is itself odorless.

Umami Basic taste (“the fifth taste”). There are two contributions to umami: a basic part, due to free glutamate, and an enhancing or synergistic part, due to the simultaneous presence of 5′-ribonucleotides, e.g., inosinate, guanylate, and adenylate.

Umibudō A green seaweed species (*Caulerpa lentillifera*) that resembles clusters of small grapes. Called green kaviar in the Philippines and sea grapes in Japan.

Uni Japanese term for sea urchin roe.

Vitellin membrane The membrane that separates the yolk from the white of an egg.

Wakame Brown seaweed (*Undaria pinnatifida*), which is widespread in Japanese cuisine.

White kaviar Popular name for white eggs from land snails.

Xanthan gum Complex polysaccharide produced by the bacterium *Xanthomonas campestris*, which can be used as a thickener.

Zona radiata (Chorion) wall of connective tissue that forms a rigid network on the inside of the outer membrane of an egg.

Bibliography

Adeli, A., & Namdar, M. (2015). The Iranian caviar and its substitutes in the world market. *Ecopersia, 3*, 933–944.

Ahn, Y.-Y., Ahnert, S. E., Bagrow, J. P., & Barabásib, A. L. (2011). Flavour network and the principles of food pairing. *Nature Scientific Reports, 1*, 196.

Arjona, O., Millan, A., Ibarra, A. M., & Palacio, E. (2008). Muscle and roe lipid composition in diploid and triploid scallops. *Journal of Food Lipids, 15*, 407–419.

Baker, A. K., Vixie, B., Rasco, B. A., Ovissipour, M., & Ross, C. F. (2014). Development of a lexicon for caviar and its usefulness for determining consumer preference. *Journal of Food Science, 79*, S2533–S2541.

Baki, B., Ozturk, D. K., & Tomgisi, S. (2021). Comparative analysis of egg biochemical composition and egg productivity rainbow trout (*Oncorhynchus mykiss* Walbaum, 1792) in different stations in Turkey. *Aquaculture Studies, 21*, 117–127.

Barrento, S., Marques, A., Teixeira, B., Mendes, R., Bandarra, N., Vaz-Pires, P., & Nunes, M. L. (2010). Chemical composition, cholesterol, fatty acid and amino acid in two populations of brown crab *Cancer pagurus*: Ecological and human health implications. *Journal of Food Composition and Analysis, 23*, 716–725.

Barrento, S., Marques, A., Teixeira, B., Mendes, R., Vaz-Pires, P., & Nunes, M. L. (2009). Nutritional quality of the edible tissues of European lobster *Homarus gammarus* and American lobster *Homarus americanus*. *Journal of Agricultural Food Chemistry, 57*, 3645–3652.

Basby, M., Jappesen, V. F., & Huss, H. H. (1998). Chemical composition of fresh and salted lumpfish (*Cyclopterus lumpus*) roe. *Journal of Aquatic Food Product Technology, 7*, 7–21.

Batista, I. (2007). By-catch, underutilized species and underutilized fish parts as food ingredients. In *Maximising the value of marine by-products* (pp. 171–195). Woodhead Publishing Series in Food Science, Technology and Nutrition.

Bekhit, A. (2022). *Fish roe: Biochemistry, products, and safety*. William Andrew Publishing.

Bekhit, A., Morton, J. D., Dawson, C. O., Zhao, J. H., & Lee, H. (2009a). Impact of maturity on the physicochemical and biochemical properties of Chinook salmon roe. *Food Chemistry, 117*, 318–325.

Bekhit, A. E., Morton, J. D., Dawson, C. O., & Sedcole, R. (2009b). Optical properties of raw and processed fish roes from six commercial New Zealand species. *Journal of Food Engineering, 91*, 363–371.

Berik, N., & Cankiriligil, C. (2013). Determination of proximate composition and sensory attributes of scallop (*Flexopecten glaber*) gonads. *Marine Science and Technology Bulletin, 2*, 5–8.

O. G. Mouritsen, K. Styrbæk, *Roe and Roe Gastronomy*,
https://doi.org/10.1007/978-3-032-13142-3

Blanchier, B., & Boucaud-Camou, E. (1984). Lipids in the digestive gland and the gonad of immature and mature *Sepia officinalis* (Mollusca: Cephalopoda). *Marine Biology, 80*, 39–43.

Blaxter, J. H. S. (1969). Development: Eggs and larvae. In W. S. Hoar & D. J. Randall (Eds.), *Fish physiology* (pp. 177–252). Academic Press.

Bledsoe, G., & Basco, B. (2006). Caviar and fish roe. In *Handbook of food science, technology, and engineering* (Vol. 4, pp. 161-1–161-20).

Bledsoe, G. E., Bledsoe, C. D., & Rasco, B. (2003). Caviars and fish roe products. *Critical Reviews in Food Science and Nutrition, 43*, 317–356.

Bowman, A. (1920). The eggs and larvae of the angler (*Lophius piscatorius*) in Scottish waters. A Review of our present knowledge of the life history of the Angler. In *Reports of the Fishery Board for Scotland Scientific Investigations for 1919*, No. II (pp. 1–42).

Bozkurt, Y., Öğetmen, F., Kökçü, Ö., & Erçin, U. (2011). Relationships between seminal plasma composition and sperm quality parameters of the *Salmo trutta macrostigma* (Dumeril, 1858) semen: With emphasis on sperm motility. *Czech Journal of Animal Science, 56*, 355–364.

Bozkurt, Y., & Seçer, S. (2006). Relationship between spermatozoa motility, egg size, fecundity and fertilization success in brown trout (*Salmo trutta fario*). *Pakistan Journal of Biological Sciences, 9*, 2141–2144.

Bronzi, P., Chebanov, M., Michaels, J. T., Wei, Q., Rosenthal, H., & Gessner, J. (2019). Sturgeon meat and caviar production: Global update 2017. *Journal of Applied Ichthyology, 35*, 257–266.

Camacho, C., Correia, T., Teixeira, B., Mendes, R., Valente, L. M. P., Pessoa, M. F., Nunes, M. L., & Gonçalves, A. (2023). Nucleotides and free amino acids in sea urchin *Paracentrotus lividus* gonads: Contributions for freshness and overall taste. *Food Chemistry, 404*, 134505.

Caprino, F., Moretti, V. M., Bellagamba, F., Turchini, G. M., Busetto, M. L., Giani, I., Paleari, M. A., & Pazzaglia, M. (2008). Fatty acid composition and volatile compounds of caviar from farmed white sturgeon (*Acipenser transmontanus*). *Analytica Chimica Acta, 617*, 139–147.

Cardinal, M., Cornet, J., & Vallet, J. L. (2002). Sensory characteristics of caviar from wild and farmed sturgeon. *International Reviews in Hydrobiology, 87*, 651–659.

Caredda, M., Addis, M., Pes, M., Fois, N., Sanna, G., Piredda, G., & Sanna, G. (2018). Physico-chemical, colorimetric, rheological parameters and chemometric discrimination of the origin of *Mugil cephalus'* roes during the manufacturing process of Bottarga. *Food Research International, 108*, 128–135.

Cejas, J. R., Almansa, E., Jérez, S., Bolaños, A., Felipe, B., & Lorenzo, A. (2004). Changes in lipid class and fatty acid composition during development in white seabream (*Diplodus sargus*) eggs and larvae. *Comparative Biochemistry and Physiology, Part B, 139*, 209–216.

Çelik, M. Y., Duman, M. B., Sariipek, M., Gören, G. U., Öztürk, D. K., Kocatepe, D., & Karayücel, S. (2019). Comparison of fatty acids and some mineral matter profiles of wild and farmed snails, *Cornu aspersum* Müller, 1774. *Molluscan Research, 39*, 234–240.

Chapman, F. A., & Van Eenennaam, J. P. (2016). Technically speaking, what is sturgeon caviar? In *Series of the school of forest resources, program in fisheries and aquatic sciences*, UF/IFAS Extension, Document FA194, 4pp.

Chen, C., Okazaki, E., & Osako, K. (2016). Textural improvement of salt-reduced Alaska pollack (*Theragra chalcogramma*) roe product by CaCl2. *Food Chemistry, 213*, 268–273.

Chiou, T. K., Matsui, T., & Konosu, S. (1989). Comparison of extractive components between raw and salted Alaska pollack roe ("Tarako"). *Nippon Suikan Gakkaishi, 55*, 115–519.

Chiou, T. K., & Konosu, S. (1988). Comparison of extractive components during processing of dried mullet roe. *Nippon Suisan Gakkaishi, 54*, 307–313.

Codex Alimentarius. (2010). *Guidelines on Nutrition Labelling CAC/GL 2–1985 as Last Amended 2010; Joint FAO/WHO*. Food Standards Programme, Secretariat of the Codex Alimentarius Commission. FAO.

Cohen, D. M., Inada, T., Iwamoto, T., & Scialabba, N. (1990). FAO species catalogue. Vol. 10. Gadiform fishes of the world (Order Gadiformes). In *An annotated and illustrated catalogue of cods, hakes, grenadiers and other gadiform fishes known to date*. FAO Fish. Synop. 125(10). 442pp. FAO.

Corrias, F., Atzei, A., Giglioli, A., Pasquini, V., Cau, A., Addis, P., Sarais, G., & Angioni, A. (2020). Influence of the technological process on the biochemical composition of fresh roe and bottarga from *Liza ramada* and *Mugil cephalus*. *Food, 9*, 1408.

Cruz-García, C., López-Hernaández, J., González-Castro, M. J., Rodíguez-Bernaldo De Quirós, A. I., & Simal-Lozano, J. (2000). Protein, amino acid and fatty acid contents in raw and canned sea urchin (*Paracentrotus lividus*) harvested in Galicia (NW Spain). *Journal of the Science of Food and Agriculture, 80*, 1189–1192.

Desnica, N., Jensen, S., Þórarinsdóttir, G. G., Jónsson, J. Ó., Kristinsson, H. G., & Gunnlaugsdóttir, H. (2011). *Icelandic blue mussels – A valuable high-quality product*. Skýrsla Matís 44–11, ISSN 1670–7192. 21pp.

Dincer, T., & Cakli, S. (2007). Chemical composition and biometrical measurements of the Turkish sea urchin (*Paracentrotus lividus*, Lamarck, 1816). *Critical Reviews in Food Science and Nutrition, 27*, 21–26.

Domínguez-Petit, R., Saborido-Rey, F., & Medina, I. (2010). Changes of proximate composition, energy storage and condition of European hake (*Merluccius merluccius*, L. 1758) through the spawning season. *Fisheries Research, 104*, 73–82.

Duarte, C. M., & Alcaraz, M. (1989). To produce many small or few large eggs: A size-independent reproductive tactic of fish. *Oecologia, 80*, 401–404.

Dunn, R., & Sanchez, M. (2021). *Delicious: The evolution of flavor and how it made us human*. Princeton University Press.

Ehrenbaum, E. (1904). Eier und Larven von Fischen der deutchen Bucht. III. Fische mit festsitzenden Eiern. *Wissenschaftliche Meeresuntersuchungen*, Abteilung Helgoland 6.

Ehrenbaum, E. (1905-1909). *Eier und Larven von Fischen des Nordischen Planktonnes*. Verlag von Lipsius & Tischer, Leipzig.

Engström, K., Wallin, R., & Saldeen, T. (2003). Effects of Scandinavian caviar paste enriched with a stable fish oil on plasma phospholipid fatty acids and lipid peroxidation. *European Journal of Clinical Nutrition, 57*, 1052–1059.

EU. (2021). *The caviar marked. Maritime Affairs and Fisheries: Production, trade, and consumption in and outside EU*. Publications Office of the European Union. ISBN 978–92–76-28914-2.

EUMOFA. (2021). *The caviar market: Production, trade, and consumption in and outside the EU*. European Marquet Observatory for Fisheries and Aquaculture Products. European Union, Publication Office of the European Union, 38pp, ISBN 978–92–76-28914-2.

EUFOMA. (2023). *Sturgeon meat and other by-products of caviar: Production, trade and consumption in and outside the EU*. European market Observatory for Fisheries and Aquaculture Products. : Publications Office of the European Union, 24pp, ISBN 978-92-76-47265-0.

Evans, R. P., Parrish, C. C., Brown, J. A., & Davis, P. J. (1996). Biochemical composition of eggs from repeat and first-time spawning captive Atlantic halibut (*Hippoglossus hippoglossus*). *Aquaculture, 139*, 139–149.

Falk-Petersen, S., Falk-Petersen, I.-B., Sargent, J. R., & Haug, T. (1986). Lipid class and fatty acid composition of eggs from the Atlantic halibut (*Hippoglossus hippoglossus*). *Aquaculture, 52*, 207–211.

FAO. (2020). *The state of world fisheries and aquaculture. Sustainability in action*. FAO.

Ferguson, J. C. (1975). The role of free amino acids in nitrogen storage during the annual cycle of a starfish. *Comparative Biochemistry and Physiology, 51A*, 341–350.

Fletcher, N. (2010). *Caviar: A global history*. Reaktion Books.

Fraser, A. J., Gamble, J. C., & Sargent, J. R. (1988). Changes in lipid content, lipid class composition and fatty acid composition of developing eggs and unfed larvae of cod (*Gadus morhua*). *Marine Biology, 99*, 307–313.

Frida. (2025). *Public food Data Base*. National Food Institute, Technical University of Denmark, https://frida.fooddata.dk/?lang=en

Frolov, A. V., & Pankov, S. L. (1992). The reproduction strategy of oyster *Ostrea edulis* L. from the biochemical point of view. *Comparative Biochemistry and Physiology, 103B*, 161–182.

Fuke, S. (1994). Taste-active components of seafoods with special reference to umami substances. In F. Shahidi & J. R. Botta (Eds.), *Seafoods: Chemistry, processing technology and quality* (pp. 115–139). Springer.

Fuke, S., & Konosu, S. (1991). Taste-active components in some foods: A review of Japanese research. *Physiology & Behavior, 49*, 863–868.

Gephart, J. A., Henriksson, P. J. G., Parker, R. W. R., Shepon, A., Gorospe, K. D., Bergman, K., Eshel, G., Golden, C. D., Halpern, B. S., Hornborg, S., Jonell, M., Metian, M., Mifflin, K., Newton, R., Tyedmers, P., Zhang, W., Ziegler, F., & Troell, M. (2021). Environmental performance of blue foods. *Nature, 597*, 360–366.

Gessner, J., Würtz, F., Kirschbaum, F., & Wirth, M. (2008). Biochemical composition of caviar as a tool to discriminate between aquaculture and wild origin. *Journal of Applied Ichthyology, 24*, 52–56.

Gong, Y., Huang, Y., Gao, L., Lu, J., Hu, Y., Xia, L., & Huang, H. (2003). Nutritional composition of caviar from three commercially farmed sturgeon species in China. *Journal of Food and Nutrition Research, 1*, 108–112.

Górka, A., Oklejewicz, B., & Duda, M. (2017). Nutrient content and antioxidant properties of eggs of the land snail *Helix aspersa maxima. Journal of Nutrition and Food Science, 7*, 1000594.

Graeve, M., & Wehrtmann, I. S. (2003). Lipid and fatty acid composition of Antarctic shrimp eggs (Decapoda: Caridea). *Polar Biology, 26*(26), 55–61.

Gunasekera, R. M., Shim, K. F., & Lam, T. J. (1996). Effect of dietary protein level on spawning performance and amino acid composition of eggs of Nile tilapia, *Oreochromis niloticus. Aquaculture, 146*, 121–134.

Hachero-Cruzado, I., Herrera, M., Quintana, D., Rodiles, A., Navas, J. I., Lorenzo, A., & Almansa, E. (2011). Total lipid and fatty acid composition of brill eggs *Scophthalmus rhombus* L. relationship between lipid composition and egg quality. *Aquaculture Research, 42*, 1011–1025.

Hachero-Cruzado, I., Herrera, M., Quintana, D., Rodiles, A., Navas, J. I., Lorenzo, A., & Almansa, E. (2013). Changes in lipid classes, fatty acids, protein and amino acids during egg development and yolk-sac larvae stage in brill (*Scophthalmus rhombus* L.). *Aquaculture Research, 44*, 1568–1577.

Hamzeh, A., Moslemi, M., Karaminasab, M., Khanlar, M. A., Faizbakhsh, R., Navai, M. B., & Tahergorabi, R. (2015). Amino acid composition of roe from wild and farmed Beluga sturgeon (*Huso huso*). *Journal of Agricultural Science and Technology, 17*, 357–364.

Harbach, H., & Palm, H. W. (2018). Development of general condition and flesh water content of long-time starved *Mytilus edulis*-like under experimental conditions. *AACL Bioflux, 11*, 301–308.

Harrison, K. E. (1990). The role of nutrition in maturation, reproduction and embryonic development of decapod crustaceans: A review. *Journal of Shellfish Research, 9*, 1–28.

Hayabuchi, H., Tou, K., Umeki, Y., Ota, A., Matsuyama, M., & Manabe, S. (2002). The relation between the brands of circulated Alaska pollack and degrees of maturity and characteristics of the eggs. *Journal of Cookery Science of Japan, 35*, 250–257.

Hayashi, T., Kohata, H., Watanabe, E., & Toyama, K. (1990). Sensory study of flavour compounds in extracts of salted salmon eggs (ikura). *Journal of the Science of Food and Agriculture, 90*, 343–356.

Hoffmann, E. (2003). Fisk lægger rigtig mange æg. *Fisk & Hav, 56*, 22–27.

İnanlı, A. G., Çoban, Ö. E., Yılmaz, Ö., Özpolat, E., & Kuzgun, N. K. (2019). Assessment of vitamin compositions and cholesterol levels of carp (*Cyprinus carpio carpio*) and rainbow trout (*Oncorhynchus mykiss*) caviars. *Ege Journal of Fisheries and Aquatic Sciences, 36*, 293–299.

Intarasirisawat, R., Benjakul, S., & Visessanguan, W. (2011). Chemical compositions of the roes from skipjack, tongol and bonito. *Food Chemistry, 124*, 1328–1334.

Itoh, K., Matsushima, H., Nozaki, Y., Osako, K., & Matubayasi, N. (2006). Studies on the composition of mullet *Mugil cephalus* roe and karasumi produced in Nagasaki. *Nippon Suisan Gakkaishi, 72*, 70–75.

Iversen, E. S. (1990). Fishing for black gold. *Sea Frontiers, 36*, 22–27.

Iwasaki, M., & Harada, R. (1985). Proximate and amino acid composition of the roe and muscle of selected marine species. *Journal of Food Science, 50*, 1585–1587.

James, P., Noble, C., Siikavuopio, S., Sloan, R., Hannon, C., Þórarinsdóttir, G., Ziemer, N., & Lochead, J. (2018). *Sea urchin fishing techniques report*. Activity A3.1.1 of the NPA URCHIN project. ISBN: 978-82-8296-372-5, 27pp.

Jarvis, N. D. (1964). Caviar and other fish roe products. U.S. Fish and Wildlife Service. *Fishery Leaflet, 567*, 1–4.

Joachim, D., & Schloss, A. (2008). *The science of good food*. Robert Rose, Inc..

Kaitaranta, J. (1980). Lipids and fatty acids of a whitefish (*Coregonus albula*) flesh and roe. *Journal of the Science of Food and Agriculture, 31*, 1301–1308.

Kaitaranta, J. K., & Ackman, R. G. (1981). Total lipids and lipid classes of fish roe. *Comparative Biochemistry and Physiology Part B: Comparative Biochemistry, 69*, 725–729.

Kaitaranta, J. K., & Linko, R. R. (1984). Fatty acids in the roe lipids of common food fishes. *Comparative Biochemistry and Physiology Part B: Biochemistry and Molecular Biology, 79*, 331–334.

Káldy, J., Mozsár, A., Fazekas, G., Farkas, M., Fazekas, D. L., Fazekas, G. L., Goda, K., Gyöngy, Z., Kovács, B., Semmens, K., Bercsényi, M., Molnár, M., & Várkonyi, E. P. (2020). Hybridization of Russian sturgeon (*Acipenser gueldenstaedtii*, Brandt and Ratzeberg, 1833) and American paddlefish (*Polyodon spathula*, Walbaum 1792) and evaluation of their progeny. *Genes, 11*, 753.

Kalogeropoulos, N., Mikellidi, A., Nomikos, T., & Chiou, A. (2012). Screening of macro- and bioactive microconstituents of commercial finfish and sea urchin eggs. *LWT - Food Science and Technology, 46*, 525/531.

Kalogeropoulos, N., Nomikos, T., Chiou, A., Fragopoulou, E., & Antonopoulos, S. (2008). Chemical composition of Greek Avgotaracho prepared from mullet (*Mugil cephalus*): Nutritional and health benefits. *Journal of Agricultural and Food Chemistry, 56*, 5916–5925.

Kasai, T. (2003). Lipid contents and fatty acid composition of total lipid of sea cucumber *Stichopus japonicus* and *konowata* (salted sea cucumber entrails). *Food Science and Technology Research, 9*, 45–48.

Kelsey, M. W. (1998). Sea-urchins: ocean hedgehogs. In H. Walker (Ed.), *Fish: Food from the Waters* (Proceedings of the Oxford Symposium on Food and Cookery 1997). Prospect Books.

Kennedy, J., Durif, C. M. F., Florin, A.-B., Freéchet, A., Gauthier, J., Hüssy, K., & S.þ. Jónsson, H.G. Ólafsson, S. Post & R.B. Hedeholm. (2019). A brief history of lumpfishing, assessment, and management across the North Atlantic. *ICES Journal of Marine Science, 76*, 181–191.

Konosu, S. (1973). Taste of fish and shellfish with special reference to taste-producing substances. *Nippon Shokuhin Kogyo Gakkaishi, 20*, 432–439.

Koshelev, V. N. (2013). Spawning migrations of Amur sturgeon *Acipenser schrenckii:* Structure of the population spawning part and gonad status of its individuals. *Journal of Ichthyology, 53*, 172–182.

Kowalska-Góralska, M., Formicki, K., Dobrzański, Z., Wondołowska-Grabowska, A., Skrzyńska, E., Korzelecka-Orkisz, A., Nędzarek, A., & Tański, A. (2020). Nutritional composition of Salmonidae and Acipenseridae fish eggs. *Annals of Animal Studies, 20*, 629–645.

Krzynowek, J., & Murphy, J. (1987). *Proximate composition, energy, fatty acid, sodium, and cholesterol content of finfish, shellfish, and their products*. NOAA Technical Report NMFS 55. U.S. Department of Commerce, National Oceanic and Atmospheric Administration, National Marine Fisheries Service. July, 53pp.

Lapa-Guimarães, J., Trattner, S., & Pickova, J. (2011). Effect of processing on amine formation and the lipid profile of cod (*Gadus morhua*) roe. *Food Chemistry, 129*, 716–723.

Lawrence, J. M. (2007). Sea urchin roe cuisine. *Developments in Aquaculture and Fisheries Science, 37*, 521–523.

Lee, C. S., Tamaru, C. D., Kelley, A. M., & Miyamoto, G. T. (1992). The effect of salinity on the induction of spawning and fertilization in the striped mullet, *Mugil cephalus*. *Aquaculture, 102*, 289–296.

Lee, J. S., Kim, J. S., Kim, J. G., Oh, K. S., Choi, B. D., Park, K. H., Choi, J. D., & J.D. (2011). Food quality characterization and safety of imported fish roe (Japanese flyingfish roe, capelin roe and Pacific herring roe). *Journal of Agriculture and Life Sciences, 45*, 95–108.

Lei, S., Zhang, X., Zhang, P., & Ikeda, Y. (2014). Biochemical composition of cuttlefish (*Sepia esculenta*) eggs during embryonic development. *Molluscan Research, 34*, 1–9.

Lopez, A., Vasconi, M., Bellagamba, F., Mentasti, T., & Moretti, V. M. (2020). Sturgeon meat and caviar quality from different cultured species. *Fishes, 5*, 9.

Lourenço, S., Valente, L. M. P., & C. Andrade C. (2019). Meta-analysis on nutrition studies modulating sea urchin roe growth, colour and taste. *Reviews in Aquaculture, 11*, 766–781.

Lourenço, S., Mendes, S., Raposo, A., Santos, P. M., Gomes, A. S., Ganhão, R., Costa, J. L., Gil, M. M., & Pombo, A. (2021). Motivation and preferences of Portuguese consumers towards sea urchin roe. *International Journal of Gastronomy and Food Science, 24*, 100312.

Lourenço, S. I., Narciso, L., Gonzalez, Á. F., Pereira, J., Auborg, S., & Xavier, J. C. (2014). Does the trophic habitat influence the biochemical quality of the gonad of *Octopus vulgaris*? Stable isotopes and lipid class contents as bio-indicators of different life cycle strategies. *Hydrobiologia, 725*, 33–46.

Lu, J. Y., Ma, Y. M., Williams, C., & Chung, R. A. (1979). Fatty acid and amino acid composition of salted mullet roe. *Journal of Food Science, 55*, 676–677.

Ma, S., Li, L. H., Hao, S. X., Yang, X. Q., Huang, H., Cen, J. W., & Wang, Y. Q. (2020). Fatty-acid profiles and fingerprints of seven types of fish roes as determined by chemometric methods. *Journal of Oleo Science, 69*, 11-99-1208.

Machado, T. M., Tabata, Y. A., Takahashi, N., & Casarini, L. M. (2016). Caviar substitute produced from roes of rainbow trout (*Oncorhynchus mykiss*). *Acta Scientiarum Technology, 38*, 233–240.

Maćkowiak-Dryka, M., Pyz-Łukasik, R., Ziomek, M., & Szkucik, K. (2020a). Nutritional value of a new type of substitute caviar. *Medycyna Weterynaryjna, 76*, 285–288.

Maćkowiak-Dryka, M., Szkucik, K., & Pyz-Łukasik, R. (2020b). Snail eggs as a raw material for the production of a caviar substitute. *Journal of Veterinary Research, 64*, 543–547.

Maćkowiak-Dryka, M., Szkucik, K., Ziomek, M., & Klimek, K. (2020c). Fatty acid profile in edible eggs of snails from the *Cornu* genus. *Journal of Veterinary Research, 64*, 137–140.

Maga, J. A. (1982). Flavour potentiators. *CRC Critical Reviews in Food Science and Nutrition, 18*, 231–312.

Margulies, D., Suter, J. M., Hunt, S. L., Olson, R. J., Scholey, V. P., Wexler, J. B., & Nakazawa, A. (2007). Spawning and early development of captive yellowfin tuna (*Thunnus albacares*). *Fishery Bulletin - National Oceanic and Atmospheric Administration, 105*, 249–265.

Massari, S., & Pastore, S. (2014). Heliciculture and snail caviar: New trends in the food sector. In M. Miśniakiewicz & S. Popek (Eds.), *Commodity Science in Research and Practice – Future trends and challenges in the food sector* (pp. 79–90. ISBN 978–83–940189-0-0). Polish Society of Commodity Science.

Mazorra, C. M., Bruce, J. G., Bell, A., Davie, E., Alorend, N., Jordan, J., Rees, N., Papanikos, M. P., & Bromage, N. (2003). Dietary lipid enhancement of broodstock reproductive performance and egg and larval quality in Atlantic halibut (*Hippoglossus hippoglossus*). *Aquaculture, 227*, 21–33.

McAlister, J. S., & Moran, A. L. (2012). Relationships among egg size, composition, and energy: A comparative study of geminate sea urchins. *PLoS One, 7*, e41599.

McGee, H. (2004). *On food and cooking: The science and Lore of the kitchen*. Scribner.

McGee, H. (2020). *Nose dive: A field guide to the world's smells*. Penguin Press.

Memmi, G. (2019). *Boutargue: Histoires, Traditions*. Recettes. Flammarion.

Méndez, E., Fermindez, M., Pazo, G., & Grompone, M. A. (1992). Hake roe lipids: Composition and changes following cooking. *Food Chemistry, 45*, 179–181.

Mol, S., & Turan, S. (2008). Comparison of proximate, fatty acid and amino acid compositions of various types of fish roes. *International Journal of Food Properties, 11*, 669–677.

Monfort, M. C. (2002). *Fish roe in Europe: Supply and demand conditions*. FAO/GLOBEFISH Research Programme; FAO: , vol. 72, 47pp.

Morais, S., Narciso, L., Calado, R., Nunes, M. L., & Rosa, R. (2002). Lipid dynamics during the embryonic development of *Plesionika martia martia* (Decapoda; Pandalidae), *Palaemon serratus* and Palaemon elegans (Decapoda; Palaemonidae): Relation to metabolic consumption. *Marine Ecology Progress Series, 242*, 195–204.

Moreno-Reyes, J. E., Mendez-Ruiz, C. A., Díaz, G. X., Meruane, J. A., & Toledo, P. H. (2015). Chemical composition of the freshwater prawn *Cryphiops caementarius* (Molina, 1782) (Decapoda: Palaemonidae) in two populations in northern Chile: Reproductive and environmental considerations. *Latin American Journal of Aquatic Research, 43*, 745–754.

Mori, M., Modena, M., & Biagi, F. (2001). Fecundity and egg volume in Norway lobster (*Nephrops norvegicus)* from different depth in the northern Tyrrhenian Sea. *Scientia Marina, 65*, 111–116.

Mouritsen, O. G. (2009). *Sushi. Food for the Eye, the Body & the Soul.* Springer, New York.

Mouritsen, O. G. (2013). *Seaweeds. Edible, Available & Sustainable.* Chicago University Press, Chicago.

Mouritsen, O. G. (2023). Roe gastronomy. *International Journal of Gastronomy and Food Science, 32*, 100712.

Mouritsen, O. G. (2024). When blue is green: Seafoods for umamification of a sustainable plant-forward diet. *International Journal of Gastronomy and Food Science, 35*, 100902.

Mouritsen, O. G., & Schmidt, C. V. (2020). A role for macroalgae and cephalopods in sustainable eating. *Frontiers in Psychology, 11*, 1402.

Mouritsen, O. G., & Styrbæk, K. (2014). *Umami. Unlocking the Secrets of the Fifth Taste.* Columbia University Press.

Mouritsen, O. G., & Styrbæk, K. (2017). *Mouthfeel: How texture makes taste*. Columbia University Press.

Mouritsen, O. G., & Styrbæk, K. (2018). *Octopuses, Squid & Cuttlefish: Seafood for today and for the future*. Springer, 2021.

Mouritsen, O. G., & Styrbæk, K. (2020). Design and 'umamification' of vegetables for sustainable eating. *International Journal of Food Design, 5*, 9–42.

Mouritsen, O. G., & Styrbæk, K. (2023). *Rogn - Meget mere end caviar*. Gyldendal.

Mouritsen, O. G. K., & Styrbæk, M. J. (2025). *Plant-forward cuisine: Basic concepts and practical applications*. Routledge.

Murillo-Navarro, R., & Jimenez-Guirado, D. (2012). Relationships between algal food and gut and gonad conditions in the Mediterranean Sea urchin *Paracentrotus lividus* (Lam.). *Mediterranean Marine Science, 13*, 227–238.

Muus, B. J., & Dahlstrøm, P. (2017). *Europas ferskvandsfisk*. Gyldendal.

Muus, B. J., & Nielsen, J. G. (2017). *Havfisk og fiskeri i Nordvesteuropa*. Gyldendal.

Nazari, R. M., Sohrabnejad, M., Ghomi, M. R., Modanloo, M., Ovissipour, M., & Kalantarian, H. (2009). Correlations between egg size and dependent variables related to larval stage in Persian sturgeon *Acipenser persicus*. *Marine and Freshwater Behaviour and Physiology, 42*, 147–155.

Nicolai, A., Filser, J., Lenz, R., Valérie, B., & Charrier, M. (2012). Composition of body storage compounds influences egg quality and reproductive investment in the land snail *Cornu aspersum*. *Canadian Journal of Zoology, 90*, 1161–1170.

Nimalaratne, C., Lopes-Lutz, D., Schieber, A., & Wu, J. (2011). Free aromatic amino acids in egg yolk show antioxidant properties. *Food Chemistry, 129*, 155–161.

Ono, M., & Mouritsen, O. G. (2025). *Traditional Japanese Flavourings and condiments: Umamification in the plant-forward cuisine*. Springer.

Ovissipour, M., & Rasco, B. (2011). Fatty acid and amino acid profiles of domestic and wild beluga (*Huso huso*) roe and impact on fertilization ratio. *Journal of Aquaculture Research and Development, 2*, 1000113.

Palacios, E., Racotta, I. S., Arjona, O., Marty, Y., Le Coz, J. R., Moal, J., & Samain, J. F. (2007). Lipid composition of the pacific lion-paw scallop, *Nodipecten subnodosus*, in relation to gametogenesis. 2. Lipid classes and sterols. *Aquaculture, 266*, 266–273.

Pandian, T. J. (1970). Ecophysiological studies on the developing eggs and embryos of the European lobster *Homarus gammarus*. *Marine Biology, 5*, 154–167.

Pappalardo, A. M., Petraccioli, A., Capriglione, T., & Ferrito, V. (2019). From fish eggs to fish name: Caviar species discrimination by Coibar-RFLP, an efficient molecular approach to detect fraud in the caviar trade. *Molecules, 24*, 2468.

Park, K. S., Kang, K. H., Bae, E. Y., Baek, K. A., Shin, M. H., Kim, D. U., Kang, H. K., Kim, K. J., Choi, Y. J., & Im, J. S. (2015). General and biochemical composition of caviar from sturgeon (*Acipenser ruthenus*) farmed in Korea. *International Food Research Journal, 22*, 777–781.

Parma, L., Bonaldo, A., Pirini, M., Viroli, C., Parmeggiani, A., Bonvini, E., & Gatta, P. P. (2015). Fatty acid composition of eggs and its relationships to egg and larval viability from domesticated common sole (*Solea solea*) breeders. *Reproduction in Domestic Animals, 50*, 186–194.

Pérez-Lloréns, J. L. (2020). Microalgae: From simple foodstuff to avant-Garde cuisine. *International Journal of Gastronomy and Food Science, 21*, 100221.

Pieters, H., Kluytmans, J. H., Zandee, D. I., & Cadée, G. C. (1980). Tissue composition and reproduction of *Mytilus edulis* in relation to food availability. *Netherlands Journal of Sea Research, 14*, 349–361.

Pleissner, D., Eriksen, N. T., Lundgreen, K., & Riisgaard, H. U. (2012). Biomass composition of blue mussels, *Mytilus edulis*, is affected by living site and species of ingested microalgae. *International Scholarly Research Network ISRN Zoology, 2012*, 902152.

Quintana, D., Márquez, L., Arévalo, J. R., Lorenzo, A., & Almansa, E. (2015). Relationships between spawn quality and biochemical composition of eggs and hatchlings of *Octopus vulgaris* under different parental diets. *Aquaculture, 446*, 206–216.

Rainis, S., Gasco, L., & Ballestrazzi, R. (2005). Comparative study on milt quality features of different finfish species. *Italian Journal of Animal Science, 4*, 355–363.

Rao, P. G. P., Balaswamy, K., Jyothirmayi, T., Karuna, M. S. L., & Prasad, R. B. N. (2015). Fish roe lipids: Composition and changes during processing and storage. *Processing and Impact on Active Components in Food, 56*, 463–468.

Rayner, T. A., Hwang, J.-S., & Hansen, B. W. (2017). Anticipating the free amino acid concentrations in newly hatched pelagic fish larvae based on recently fertilized eggs and temperature. *Journal of Plankton Research, 39*, 1012–1019.

Ren, J. S., Marsden, I. D., Ross, A. H., & Schiel, D. R. (2003). Seasonal variation in the reproductive activity and biochemical composition of the Pacific oyster (*Crassostrea gigas*) from the Marlborough sounds, New Zealand. *New Zealand Journal of Marine and Freshwater Research, 37*, 171–182.

Riis-Vestergaard, J. (1982). Water and salt balance of halibut eggs and larvae (*Hippoglossus hippoglossus*). *Marine Biology, 70*, 135–139.

Rosa, R., Costa, P. R., & Nunes, M. L. (2004a). Effect of sexual maturation on the tissue biochemical composition of *Octopus vulgaris* and *O. Defilippi* (Mollusca: Cephalopoda). *Marine Biology, 145*, 563–574.

Rosa, R., Costa, P. R., Pereira, J., & Nunes, M. L. (2004b). Biochemical dynamics of spermatogenesis and oogenesis in *Eledone cirrhosa* and *Eledone moschata* (Cephalopoda: Octopoda). *Comparative Biochemistry and Physiology, Part B, 139*, 299–310.

Rosa, R., & Nunes, M. L. (2003). Nutritional quality of red shrimp, *Aristeus antennatus* (Risso), pink shrimp, *Parapenaeus longirostris* (Lucas), and Norway lobster, *Nephrops norvegicus* (Linnaeus). *Journal of the Science of Food and Agriculture, 84*, 89–94.

Rosa, R., Nunes, L., & Reis, C. S. (2002). Seasonal changes in the biochemical composition of *Octopus vulgaris* Cuvier, 1797, from three areas of the Portuguese coast. *Bulletin of Marine Science, 71*, 739–751.

Rosa, R., Morais, S., Calado, R., Narciso, L., & Nunes, M. L. (2003). Biochemical changes during the embryonic development of Norway lobster, *Nephrops norvegicus*. *Aquaculture, 221*, 507–522.

Rosa, A., Scano, P., Melis, M.-P., Deiana, M., Atzeria, A., & Dessì, M. A. (2009). Oxidative stability of lipid components of mullet (*Mugil cephalus*) roe and its product "bottarga". *Food Chemistry, 115*, 891–896.

Rygało-Galewska, Zgliínska, K., & Niemiec, T. (2022). Edible snail production in Europe. *Animals, 12*, 2732.

Rønnestad, I., Groot, E. P., & Fyhn, H. J. (1993). Compartmental distribution of free amino acids and protein in developing yolk-sac larvae of Atlantic halibut (*Hippoglossus hippoglossus*). *Marine Biology, 116*, 349–354.

Rønnestad, I., Fyhn, H. J., & Gravningen, K. (1992). The importance of free amino acids to the energy metabolism of eggs and larvae of turbot (*Scophthalmus maximus*). *Marine Biology, 114*, 517–525.

Rønnestad, I., Thorsen, A., & Finn, R. N. (1999). Fish larval nutrition: A review of recent advances in the roles of amino acids. *Aquaculture, 177*, 201–216.

Rønnestad, I., Robertson, R. R., & Fyhn, H. J. (1996). Free amino acids and protein content in pelagic and demersal eggs of tropical marine fishes. In D. D. MacKinlay & M. Eldridge (Eds.), *The fish egg* (pp. 81–84). American Fisheries Society, Physiology Section.

Saffron, I. (2002). *Caviar – The strange history and uncertain future of the world's Most coveted delicacy*. Broadway Books.

Salman, Y., Salman, A., & Ozkizilcik, S. (2007). The fatty acid profile of the marine cephalopod *Loligo vulgaris*. *The Israeli Journal of Aquaculture – Bamidgeh, 59*, 133–136.

Salonen, E. (2003). Here we come, world! *Gastronomica–The Journal of Food and Culture, 3*, 101–103.

Salze, G., Tocher, D. R., William, J. R., & Robertson, D. A. (2005). Egg quality determinants in cod (*Gadus morhua* L.): Egg performance and lipids in eggs from farmed and wild broodstock. *Aquaculture Research, 36*, 1488–1499.

Sanchez-Velasco, L., & Norbis, W. (1997). Comparative diets and feeding habits of *Boops boops* and *Diplodus sargus* larvae, two sparid fishes co-occurring in the northwestern Mediterranean (May 1992). *Bulletin of Marine Science, 61*, 821–835.

Sarower, M. G., Hasanuzzaman, A. F. M., Biswas, B., & Abe, H. (2012). Taste producing components in fish and fisheries products: A review. *International Journal of Food Fermentation Technology, 2*, 113–121.

Scano, P., Rosa, A., Marincola, F. C., Locci, E., Melis, M. P., Dessì, M. A., & Laia, A. (2008). 13C NMR, GC and HPLC characterization of lipid components of the salted and dried mullet (*Mugil cephalus*) roe "bottarga". *Chemistry and Physics of Lipids, 151*, 69–76.

Scheibling, R. E., & Lawrence, J. M. (1982). Differences in reproductive strategies of morphs of the genus *Echinaster* (Echinodermata: Asteroidea) from the eastern Gulf of Mexico. *Marine Biology, 70*, 51–62.

Schmidt, C. V., & Mouritsen, O. G. (2020). The solution to sustainable eating is not a one-way street. *Frontiers in Psychology, 11*, 531.

Schmidt, C. V., & Mouritsen, O. G. (2022). Umami taste as a driver for sustainable eating. *International Journal of Food Design, 7*, 187–203.

Schmidt, C. V., Olsen, K., & Mouritsen, O. G. (2020). Umami synergy as the scientific principle behind taste-pairing champagne and oysters. *Nature Scientific Reports, 10*, 20077.

Schmidt, C. V., Raza, H., Olsen, K., & Mouritsen, O. G. (2024). Proximate nutritional composition of roe from fish, crustaceans, mussels, echinoderms, and cephalopods. *International Journal of Gastronomy and Food Science, 36*, 100944.

Schrader, K. K., & Rimando, A. M. (2003). *Off-flavors in aquaculture: An overview in off-flavors in aquaculture* (Vol. 848, pp. 1–12). American Chemical Society.

Schram, E., Kwadijk, C., Hofman, A., Blanco, A., Murk, A., Verreth, J., & Schrama, J. (2021). Effect of feeding during off-flavour depuration on geosmin excretion by Nile tilapia (*Oreochromis niloticus*). *Aquaculture, 531*, 735883.

Schubring, R. (2004). Differential scanning calorimetric (DSC) measurements on the roe of rainbow trout (*Oncorhynchus mykiss*): Influence of maturation and technological treatment. *Thermochimica Acta, 415*, 89–98.

Şen, H. (2005). Temperature tolerance of Loliginid squid (*Loligo vulgaris* Lamarck, 1798) eggs in controlled conditions. *Turkish Journal of Fisheries and Aquatic Sciences, 5*, 53–56.

Serezli, R., Güzel, Ş., & Kocabaş, M. (2010). Fecundity and egg size of three salmonid species (*Oncorhynchus mykiss*, *Salmo labrax*, *Salvelinus fontinalis*) cultured at the same farm condition in north-eastern, Turkey. *Journal of Animal and Veterinary Advances, 9*, 576–580.

Sewell, M. A. (2005). Utilization of lipids during early development of the sea urchin *Evechinus chloroticus*. *Marine Ecology Progress Series, 304*, 133–142.

Shahidi, F., Metusalach, & Brown, J. A. (1998). Carotenoid pigments in seafoods and aquaculture. *Critical Reviews in Food Science, 38*, 1–67.

Shimada, K., & Ogura, N. (1990). Lipid changes in sea urchin gonads during storage. *Journal of Food Science, 55*, 967–971.

Shirai, N., Higuchi, T., & Suzuki, H. (2006). Analysis of lipid classes and the fatty acid composition of the salted fish roe food products, Ikura, Tarako, Tobiko and Kazunoko. *Food Chemistry, 94*, 61–67.

Shuang, M., Shuxian, H., Laihao, L., Xianqing, Y., Hui, H., Ya, W., Jianwei, C., & Honglei, Z. I. (2019). Comparative analysis of nutritional components of several roes. *South China Fisheries Science, 15*, 113–121.

Sicuro, B. (2019). The future of caviar production on the light of social changes: A new dawn for caviar? *Reviews in Aquaculture, 11*, 204–219.

Siddique, M. A. M., Cosson, J., Psenicka, M., & Linhart, O. (2014). A review of the structure of sturgeon egg membranes and of the associated terminology. *Journal of Applied Ichthyology, 30*, 1246–1255.

Simopoulos, A. P. (2016). An increase in the omega-6/omega-3 fatty acid ratio increases the risk for obesity. *Nutrients, 8*, 128.

Smith, J. G., Tomoleoni, J., Staedler, M., Lyon, S., Fujii, J., & Tinker, M. T. (2021). Behavioral responses across a mosaic of ecosystem states restructure a sea otter–urchin trophic cascade. *Proceedings of the National Academy of Sciences USA, 118*, e2012493118.

Soudant, P., Marty, Y., Moal, J., Robert, R., Quéré, C., Le Coz, J. R., & Samain, J. F. (1996). Effect of food fatty acid and sterol quality on Pecten maximus gonad composition and reproduction process. *Aquaculture, 143*, 361–378.

Soudant, P., Van Ryckeghem, K., Marty, Y., Moal, J., Samain, J. F., & Sorgeloos, P. (1999). Comparison of the lipid class and fatty acid composition between a reproductive cycle in nature and a standard hatchery conditioning of the Pacific oyster *Crassostrea gigas*. *Comparative Biochemistry and Physiology Part B, 123*, 209–222.

Soundarapandian, P., & Singh, R. K. (2008). Biochemical composition of the eggs of commercially important crab *Portunus pelagicus* (Linnaeus). *International Journal of Zoological Research, 4*, 53–58.

Stefánsson, G., Kristinsson, H., Ziemer, N., Hannon, C., & James, P. (2017). *Markets for sea urchins: A review of global supply and markets* (pp. 10–17, ISSN 1670–7192, 44pp). Skýrsla Matís.

Styrbæk, K., & Mouritsen, O. G. (2021). *Grønt med umami og velsmag. Håndværk, viden & opskrifter.* Gyldendal.

Suzuki, T., & Suyama, M. (1983). Free amino acids and phosphopeptides in the extracts of fish eggs. *Bulletin of the Japanese Society of Scientific Fisheries, 49*, 1747–1753.

Sykes, A. V., Almansa, E., Lorenzo, A., & Andrade, J. P. (2009). Lipid characterization of both wild and cultured eggs of cuttlefish (*Sepia officinalis* L.) throughout the embryonic development. *Aquaculture Nutrition, 15*, 38–53.

Tagaki, T., Eaton, C. A., & Ackman, R. G. (1985). Distribution of fatty acids in lipids of the common Atlantic Sea urchin *Strongylocentrotus droebachiensis*. *Canadian Journal of Fisheries and Aquatic Sciences, 37*, 195–202.

Tanikawa, E., Motohiro, T., & Akiba, M. (1985). *Marine products in Japan* (pp. 193–240). Koseisha Koseikaku Co., Ltd..

Tavakoli, S., Luo, Y., Regenstein, J. M., Daneshvar, E., Bhatnagar, A., Tan, Y., & Hong, H. (2021). Sturgeon, caviar, and caviar substitutes: From production, gastronomy, nutrition, and quality change to trade and commercial mimicry. *Reviews in Fisheries Science & Aquaculture, 29*, 753–768.

Tocher, D. R., & Sargent, J. R. (1984). Analyses of lipids and fatty acids in ripe roes of some northwest European marine fish. *Lipids, 19*, 492–499.

Tsuyuki, H., & Fuke, S. (1978). *Alaskan pollock roe processing: A description of current Japanese industrial methods and their adaptation to the fishery in British Columbia*. Technology Services Branch Fisheries Management, Pacific Region Fisheries and Oceans Canada, Minister of Supply and Services Canada, Cat. no. Fs 97–6/1978–851, ISSN 0701–7676. Fisheries and Marine Service Technical Report 851, 28pp.

Tuck, I. D., Taylor, A. C., Atkinson, R. J. A., Gramitto, M. E., & Smith, C. (1997). Biochemical composition of *Nephrops norvegicus*: Changes associated with ovary maturation. *Marine Biology, 129*, 505–511.

Turchini, G., Torstensen, B. E., & Ng, W. K. (2009). Fish oil replacement in finfish nutrition. *Reviews in Aquaculture, 1*, 10–57.

Ueda, R., Okamoto, N., Araki, T., Shibata, M., Sagara, Y., Sugiyama, K., & Chiba, S. (2009). Consumer preference and optical and sensory properties of fresh cod roe. *Food Science and Technology Research, 15*, 469–478.

UIC. (2025). *Umami Information Center*. www.umamiinfo.com

Valerón, N. R., Vásquez, D. P., Rodgers, R., Rugset, K. O., & Munk, R. (2022). From waste to a new sustainable ingredient in the kitchen: Red king crab abdominal flap (*Paralithodes camtschaticus*). *International Journal of Gastronomy and Food Science, 27*, 100455.

Valverde, C. J., Martínez-Llorens, S., Vidal, A. T., Cerda, M. J., Rodriguez, C., Estefanell, J., Gairin, J. I., Domingues, P. M., Rodríguez, C. J., & Garcia, B. G. (2013). Amino acids composition and protein quality evaluation of marine species and meals for feed formulations in cephalopods. *Aquaculture International, 21*, 413–433.

Van Uhm, D., & Siegel, D. (2016). The illegal trade in black caviar. *Trends in Organized Crime, 19*, 67–87.

Vasconi, M., Tirloni, E., Stella, S., Coppola, C., Lopez, A., Bellagamba, F., Bernardi, C., & Moretti, V. M. (2020). Comparison of chemical composition and safety issues in fish roe products: Application of chemometrics to chemical data. *Food, 9*, 540.

Vega, C., & Mercadé-Prieto, R. (2011). Culinary biophysics: On the nature of the 6X°C egg. *Food Biophysics, 6*, 152–159.

Vilgis, T. A. (2020). The physics of the mouthfeel of caviar and other fish roe. *International Journal of Gastronomy and Food Science, 19*, 100192.

Vúorela, R., Kaitaranta, J., & Linko, R. R. (1979). Proximate composition of fish roe in relation to maturity. *Canadian Institute of Food Technology Journal, 12*, 186–188.

Wehrtmann, I. S., & Graeve, M. (1998). Lipid composition and utilization in developing eggs of two tropical marine caridean shrimps (Decapoda: Caridea: Alpheidae, Palaemonidae). *Comparative Biochemistry and Physiology Part B, 121*, 457–463.

Willett, W., Rockström, J., Loken, B., Springmann, M., Lang, T., Vermeulen, S., Garrett, T., Tilman, D., DeClerk, F., Wood, A., Jonell, M., Clark, M., Gordon, L. J., Fanzo, J., Hawkes, C., Zurayk, R., Rivera, J. A., De Vries, W., Sibanda, L. M., Afshin, A., Chaudhary, A., Herrero, M., Agustina, R., Branca, F., Lartey, A., Fan, S., Crona, B., Fox, E., Bignet, V., Troell, M., Lindahl, T., Singh, S., Cornell, S. E., Reddy, K. S., Narain, S., Nishtar, S., & Murray, C. J. L. (2019). Food in the Anthropocene: The EAT–lancet commission on healthy diets from sustainable food systems. *Lancet, 393*, 447–492.

White, A., & Fletcher, T. C. (1987). Polar and neutral lipid composition of the gonads and serum of the plaice, *Pleuronectes platessa* L. *Fish Physiology and Biochemistry, 4*, 37–43.

Wirth, M., Kirschbaum, F., Gessner, J., Williot, P., Patriche, N., & Billard, R. (2002). Fatty acid composition in sturgeon caviar from different species: Comparing wild and farmed origins. *International Review of Hydrobiology, 87*, 629–636.

Wong, C., & Thuras, D. (2021). *Gastro Obscura: A food adventurer's guide*. Workman Publishing.

Wrangham, R. (2009). *Catching fire: How cooking made us human*. Basic Books.

Yamaguchi, S., & Kimizuka, A. (1979). Psychometric studies on the taste of monosodium glutamate. In L. J. Filer, S. Garattini, M. R. Kare, W. A. Reynolds, & R. J. Wurtman (Eds.), *Glutamic acid: Advances in Biochemistry and Physiology* (pp. 35–54). Raven Press.

Yamaguchi, S., Yoshikawa, T., & S., Ikeda & T. Ninomiya, T. (1971). Measurement of the relative taste intensity of some L-α-amino acids and 5′-nucleotides. *Journal of Food Science, 36*, 846–849.

Yeşilayer, N., & Türk, E. (2018). Determination of fatty acid compositions in eggs of commercial aquafeed fed broodstocks rainbow trout (*Oncorhynchus mykiss*) and brown trout (*Salmo trutta* spp.). *Journal of New Results in Science, 7*, 67–75.

Yoon, I. S., Lee, G. W., Kang, S. I., Park, S. Y., Lee, J. S., Kim, J. S., & Heu, M. S. (2018). Chemical composition and functional properties of roe concentrates from skipjack tuna (*Katsuwonus pelamis*) by cook-dried process. *Food Science and Nutrition, 6*, 1276–1286.

Yu, H., Li, R., Liu, S., Xing, R., Chen, X., & Li, P. (2014). Amino acid composition and nutritional quality of gonad from jellyfish *Rhopilema esculentum. Biomedicine & Preventive Nutrition, 4*, 399–402.

Zabyelina, Y. G. (2014). The "fishy" business: A qualitative analysis of the illicit market in black caviar. *Trends in Organized Crime, 17*, 181–198.

Zagalsky, P. F. (1985). A study of the astaxanthin-lipovitellin, ovoverdin, isolated from the ovaries of the lobster, *Homarus gammarus* (L.). *Comparative Biochemistry and* Physiology *Part B: Comparative Biochemistry, 80*, 589–597.

Scientific Index

O. G. Mouritsen, K. Styrbæk, *Roe and Roe Gastronomy*,
https://doi.org/10.1007/978-3-032-13142-3

Subject Index

O. G. Mouritsen, K. Styrbæk, *Roe and Roe Gastronomy*,
https://doi.org/10.1007/978-3-032-13142-3

T